Lecture Notes in Physics

Edited by H. Araki, Kyoto, J. Ehlers, München, K. Hepp, Zürich
R. Kippenhahn, München, D. Ruelle, Bures-sur-Yvette
H. A. Weidenmüller, Heidelberg, J. Wess, Karlsruhe and J. Zittartz, Köln
Managing Editor: W. Beiglböck

349

Y. A. Kubyshin J. M. Mourão
G. Rudolph I. P. Volobujev

Dimensional Reduction of Gauge Theories, Spontaneous Compactification and Model Building

Springer-Verlag
Berlin Heidelberg GmbH

Authors

Yura A. Kubyshin
Igor P. Volobujev
Nuclear Physics Institute, Moscow State University
Moscow 119899, USSR

José M. Mourão
CFN, University of Lisboa
Av. Gama Pinto 2, P-1699 Lisboa Codex, Portugal

Gerd Rudolph
NTZ and Sektion Physik, KMU Leipzig
Karl-Marx-Platz 10/11, 7010 Leipzig, GDR

ISBN 978-3-662-13753-6 ISBN 978-3-540-46860-8 (eBook)
DOI 10.1007/978-3-540-46860-8

Originally published by Springer-Verlag Berlin Heidelberg New York in 1989

2153/3140-543210 – Printed on acid-free paper

Abstract: In the first part dimensional reduction of pure Yang-Mills theories is discussed; in particular, a general method for calculating the scalar field potential is presented. In the second part, dimensional reduction of gravity (including torsion) and spontaneous compactification within Einstein-Yang-Mills systems are considered. Using the dimensional reduction method, we present a general procedure for solving the equations of spontaneous compactification. Finally, dimensional reduction of matter fields (including fermions) is discussed, and comments on attempts to construct realistic models are made. A few illustrative examples are given.

Contents

List of important notations

Let M_1 and M_2 be smooth manifolds and $f: M_1 \longrightarrow M_2$ be a smooth mapping. We denote the tangent mapping of f at $x \in M_1$ by $f_x^{`}$, $f_x^{`}: T_x M_1 \longrightarrow T_{f(x)} M_2$. The pullback of a differential form ω via f will be denoted by $f^* \omega$.

E - multidimensional universe, (smooth, orientable manifold, $\dim E = D$)

M - 4-dimensional space time, (smooth, orientable manifold), immersed space time is denoted by $\tilde{M} \equiv s(M)$, $s: M \longrightarrow E$

I - internal space, (smooth, orientable, compact manifold, $\dim I = d$)

K - gauge group, (compact, connected Lie group)

$\mathfrak{k}$ - Lie algebra of K

Ad - adjoint representation of a Lie group

ad - adjoint representation of a Lie algebra

$\hat{P}(E,K,\pi,\psi)$ - principal fibre bundle over E with structure group K, canonical projection π and right action of K denoted by ψ (by reduction obtained subbundles are denoted by $\tilde{P}, P$ ...)

$\psi_{\hat{p}}^{`}(k)$ - Killing field (at $\hat{p} \in \hat{P}$) of the right action of K, generated by $k \in \mathfrak{k}$

$Aut(\hat{P})$ - group of automorphisms of $\hat{P}$

$Aut_o(\hat{P})$ - group of vertical automorphisms of $\hat{P}$

$\hat{\omega}$ - gauge potential (connection form at $\hat{P}$), by reduction obtained connection forms in $\tilde{P}, P$ ... are denoted by $\tilde{\omega}, \omega$ ..., their local representatives are denoted by $\hat{A}, \tilde{A}, A$, ...

$\hat{\Omega}$ - curvature form of $\hat{\omega}$, the local representative is denoted by $\hat{F}$

ϕ, Ψ - matter fields, denote (in different chapters different) sections of bundles associated with a principal bundle $\hat{P}$ (often considered as equivariant mappings from P into the typical fibre of the associated bundle), also we use $\hat{\phi}$, $\hat{\Psi}$, $\tilde{\phi}$, $\tilde{\Psi}$, ...

$D\phi$ - covariant derivative of ϕ with respect to a given connection form

G - symmetry group, acting on E, $\delta : G \times E \longrightarrow E$, (compact connected Lie group)

$\mathfrak{g}$ - Lie algebra of G

$\hat{\delta}$ - lift of the G-action δ to automorphisms of $\hat{P}$

$\hat{\delta}_{\hat{p}}(u)$- Killing field (at $\hat{p} \in \hat{P}$) of the lifted G-action, generated by $u \in \mathfrak{g}$

H - stabilizer of the G-action on E

$\mathfrak{h}$ - Lie algebra of H

τ - homomorphism of H into K

N - normalizer of H in G

$\mathfrak{n}$ - Lie algebra of N/H

C - centralizer of $\tau(H)$ in K

$\mathfrak{c}$ - Lie algebra of C

$\mathfrak{m}$ - denotes (in different chapters different) vector subspaces of Lie algebras

$\langle \cdot , \cdot \rangle_{\mathfrak{g}(\mathfrak{k})}$ - Ad-invariant scalar product in $\mathfrak{g}$, resp. $\mathfrak{k}$

W(M,N/H) - principal bundle over M with structure group N/H

$\hat{\gamma}$ - metric on E, often treated as a mapping $\hat{\gamma} : TE \longrightarrow T^*E$

γ - G-invariant metric on G/H

$\tilde{\gamma}$ - induced (by $\hat{\gamma}$) metric on $\tilde{M}$, (resp. its pull-back to M)

η - Minkowski metric

ξ - connection form on W(M,N/H)

Ξ - curvature form of ξ

$*$ - Hodge-star operation, associated with $\hat{\gamma}$

v_E - volume form on E, (corresponding notation for volume forms on other manifolds)

$(\cdot , \cdot)_{\hat{p}}$ - scalar product in the space of horizontal forms on $\hat{P}$ (at $\hat{p}$), with values in a vector space vol(·) - volume of the group (·)

V(ϕ) - scalar field potential

Θ - canonical, Lie algebra valued 1-form on a Lie group

R - gauge group after spontaneous symmetry breaking

$\mathcal{R}$ - Lie algebra of R

$\mathfrak{z}_i$ - Abelian ideals in $\mathfrak{h}$

$\mathfrak{h}_i$ - non-Abelian ideals in $\mathfrak{h}$

$\alpha = [\alpha^{(1)}, \dots, \alpha^{(p)}](\varkappa^{(1)}, \dots, \varkappa^{(q)})$ - characterization of irreducible representations of $\mathfrak{h} = \mathfrak{h}_1 \oplus \dots \oplus \mathfrak{h}_p \oplus \mathfrak{z}_1 \oplus \dots \oplus \mathfrak{z}_q$

$E^{(k)s}{}_i$ - basic intertwining operators

m_k - parameters, characterizing the geometry of G/H

$L^H(\mathfrak{m}, \mathfrak{k})$ - space of matter fields fulfilling constraint equation (1.3.3)

$\|\cdot\|$ - norm in $L^H(\mathfrak{m}, \mathfrak{k})$, norms in subspaces are denoted by $\|\cdot\|_k$

L(E) - bundle of linear frames over E, $e \in L(E)$

$\hat{O}(E)$ - bundle of orthonormal frames on E

$\tilde{O}(E)$ - bundle of adapted orthonormal frames on E

O(E) - bundle of adapted orthonormal frames on E with fixed component in orbit direction

$\mathcal{V}$ - vertical distribution on E, spanned by Killing fields of the group action

$\mathcal{H}$ - horizontal (with respect to ξ) distribution on E

$\hat{\mathcal{V}}(\hat{\mathcal{H}})$ - lift of $\mathcal{V}(\mathcal{H})$ to O(E)

$\hat{\varphi}$ - contorsion form on $\hat{O}(E)$

φ - contorsion form on O(M)

$\hat{\mathring{\omega}}$ - Levi-Civita connection form corresponding to $\hat{\gamma}$

$\mathring{\omega}$ - Levi-Civita connection form correoponding to $\tilde{\gamma}$

$\hat{\vartheta}$ - canonical R^D-valued 1-form on $\hat{O}(E)$

ϑ - canonical R^4-valued 1-form on O(M)

$\langle\cdot\,,\cdot\rangle$ - standard scalar product on R^D, associated with η

$\hat{R}$ - scalar curvature on E

R_M - scalar curvature on M

∇ - full covariant derivative (with respect to ξ and $\mathring{\omega}$ o(1,3))

h_{ab} - field with values in scalar products on $\mathfrak{g}$

$\hat{Q}$ - sum of $\hat{O}(E)$ and $\hat{P}(E,K)$ with structure group $\hat{S} = O(1,D-1) \times K$

$\hat{\varphi}$ - matter field of arbitrary spin-tensor type, $\hat{\varphi} : \hat{Q} \longrightarrow \mathcal{R}_D \equiv \mathcal{R}_\Delta \otimes \mathcal{R}_\alpha$

Q - sum of bundles $O(\tilde{M})$ and $P(\tilde{M},C)$, with structure group $S = O(1,3) \times C$

$\hat{\Gamma}^A$ - Dirac matrices on E

Indices A,B, ... resp. M,N, ... refer to objects on R^D resp. E, α, β, ... are indices in R^d, a,b,c, indices of objects on G/H resp. $\mathfrak{g}$ and μ, ν, ... refer to objects on M resp. R^4.

0. Introduction

Nowadays almost all physicists working in the field of quantum field theory and particle physics agree that the Weinberg-Salam-Glashow model and quantum chromodynamics describe the electroweak and strong interactions correctly, at least at energies of present day colliders. Both of these theories are gauge theories, and their experimental confirmation, in particular the discovery of the intermediate vector bosons, was a strong argument supporting the attitude that the fundamental physical principle for the description of the interaction of elementary particles is that of local gauge invariance.

Now we are challenged to find a theory unifying all fundamental interactions. Such a theory should be - at least in its low-energy-limit - a gauge theory, it should explain chirality and the number of fermion families, the origin of Higgs fields and it should contain only a small number of parameters, playing the role of fundamental constants.

One approach to realize such a programme is based on gauge theories of grand unification type on 4-dimensional space time [1] . Further development of this direction is connected with supersymmetric theories [2] and, in particular, with supergravity [3] . Another approach is based on considering theories on multidimensional universes - space time manifolds of dimension greater than four. This approach is closely connected with the old ideas of Kaluza and Klein [4], see also [5]. However, in recent years the development of the subject led to an intersection of these two lines of research. At one hand, $N \geq 4$ supergravity [3,6] and the theory of superstrings [7] naturally live on spaces of dimension greater than four. On the other hand, there are attempts to construct grand unification models in four dimensions using the dimensional reduction method, see section 3.4 .

In a modern version, the Kaluza-Klein idea can be described as follows [6,8-11]. One starts with a theory of either the gravitational field, or the gravitational field interacting with bosonic matter fields on a manifold E, $\dim E = D > 4$. In a first step one looks for solutions to the field equations on E, corresponding to a factorization E = MxI, with M denoting 4-dimensional space time and I being an internal space, usually assumed to be space-like and compact. Such a solution is called vacuum. If one starts with pure gravity, one cannot get interesting solutions of the form E = MxI; in this case the only possibilities for I are tori or compact manifolds with no symmetries at all, e.g. K3-spaces [6]. Thus, one has to add bosonic matter. Typically, one adds Yang-Mills fields [12-14], see also [15-22] . Of course, in doing so, one partially departs from the original Kaluza-Klein idea. On the other hand, in supergravity theories on multidimensional universes bosonic matter fields appear naturally. The typical example is D=11 supergravity, admitting the famous Freund-Rubin solutions [23]. Moreover, as argued in [24], the existence of topologically nontrivial gauge field (vacuum) configurations - like monopoles and instantons - on the internal space may be helpful in obtaining chiral fermions in four dimensions, see also [25,26]. Depending on the choice of matter fields on the multidimensional level, one can get different types of internal spaces [27]. For example in the heterotic string theory with gauge group E_8xE_8 , the most popular compactification mechanism leads to Calabi-Yau-manifolds or orbifolds [28]. If one allows fermion condensates to participate in the compactification mechanism, then one obtains certain types of homogeneous spaces for I [29]. In this Review we restrict ourselves to the case of homogeneous internal spaces, I = G/H. These spaces and, in particular, the special classes of symmetric and isotropy-irreducible spaces are exhaustively investigated in the mathematical literature [30,31]. Therefore, in this case a systematic study of the dimensional reduction procedure is possible. Closing our digression on the multidimensional vacuum, we mention non-standard approaches in which either the internal space contains time like directions [16,32] , or is non-compact (but with small volume) [33].

Now, if a vacuum solution has been found, in a second step one has to find an effective theory in four dimensions. For that purpose one considers small fluctuations about the ground state values of the D-dimensional fields, linearizes the field equations with respect to these fluctuations and expands them harmonically on G/H. Finally, one integrates over G/H. The result is an effective 4-dimensional theory containing an infinite tower of massive states with masses of the order of the Planck mass and a finite number of massless states. At present time there does not exist a constructive method to investigate the effective theory including the whole infinite tower of states. Therefore, one usually restricts oneself to a certain sector.

One approach [6,35] consists in discarding all massive states, saying that particles with masses of the order of the Planck mass are not observable.The observable particles ("low energy sector") are supposed to correspond to the massless states. Finally, one hopes that the massless fields obtain their small masses via quantum effects. However, it turns out that after discarding the massive modes, consistency (on the level of the field equations) of the effective 4-dimensional theory with the original multidimensional theory is lost [36] . This seems to be a serious disadvantage of this approach. A consistent effective theory in four dimensions is obtained, if one restricts oneself to the sector of G-symmetric modes [36]. This sector usually does not coincide with the sector of massless modes [37]; it also contains massive fields. But, both types of fields are different manifestations of one multidimensional symmetric field, and their interaction and dynamics are governed by the field equations of the symmetric sector.

It is the symmetric sector which we discuss in this Review. If one considers pure Yang-Mills theories on multidimensional universes of the form $E = M \times G/H$, then the procedure of finding the effective 4-dimensional theory corresponding to the symmetric sector is called Coset Space Dimensional Reduction (CSDR) . This special type of a multidimensional theory, discussed in detail in chapter 1, was inspired by investigations on special solutions of the Yang-Mills equations exhibiting space time symmetries. It was Witten [38] who observed that the reduction of an SU(2)-Yang-Mills theory with spherically symmetric potentials in four dimensions leads to a Higgs model in two dimensions. In a next step the notion of a spherically symmetric gauge potential was generalized to the notion of a G-symmetric potential [39,40] ; in particular, the relation with the notion of a G-invariant connection was observed. A first systematic study of the reduction of the physical action for arbitrary G-symmetric gauge potentials was presented in [41]. Using methods of local differential geometry (Lie-derivative-techniques) a classification of G-symmetric gauge potentials was given. Moreover, for the case of a metric on MxG/H of the form $\check{\gamma} \oplus \gamma$, with γ being G-symmetric, the reduction of the pure Yang-Mills action was performed. The result was a Yang-Mills-Higg system in four dimensions with a quartic self-interaction potential. It was also shown that the Higgs fields had to fulfill an algebraic constraint equation, inherited from G-symmetry of the multidimensional gauge potential, which was interpreted in group theoretical terms. In general, finding the explicit form of the potential amounted to solving this constraint equation.

The problems discussed in [41] were reconsidered in terms of global differential geometry (fibre bundle methods) in [42-47], and - in particular - the considerations were extended to the case of arbitrary G-symmetric metrics on E, see [45,46]. In [43], there was pointed out a special class of models, namely such for which the symmetry group G has an isomorphic image in the gauge group K. In this case the constraint equation for the Higgs field is solved trivially and spontaneous symmetry breaking induced by the Higgs potential can be discussed easily. Therefore, this class of models has been prefered by model builders for a long time. On the other hand, it was noticed that this class of models has some serious disadvantages, see subsection 2.3.3 and section 3.3.

Systematic studies of the group-theoretical aspect of the CSDR-scheme in more general cases were presented in [48-51] . In particular, a complete analysis of the structure of the scalar field potential for the case of G/H being symmetric has been given [50]. In [25,52-59] it was shown how to include fermions; in particular, the problem of obtaining chiral asymmetry in four dimensions was studied [25,53,59] , see sections 3.1 and 3.3. There has been a lot of attempts to use the CSDR-scheme for the construction of realistic models, e.g. the bosonic sector of the Weinberg-Salam model [60,49,53] or some variants of grand unified models [52,61-67], see chapter 3. The most attractive feature of this type of model building is that one has only a small number of free parameters, the original gauge coupling constant and one - or in more complicated cases - a few parameters characterizing the size of the internal space. This allows to predict mixing angles and masses of the Higgs bosons.

Another line of research is the investigation of the symmetric sector of pure gravity on multidimensional universes. The case, when the internal space is a non-Abelian Lie group has been studied in [68-70]; a generalization to the case of homogeneous spaces was given in [71]. If one performs the dimensional reduction of the pure Einstein-Hilbert action on E, then one obtains an effective theory in four dimensions, describing the interaction of gravity with a Yang-Mills field and a set of scalar fields. However, the scalar fields are non-minimally coupled to the Yang-Mills field and the self-interaction scalar field potential is usually not of Higgs type. Thus, pure gravitational theories on multidimensional universes cannot be used to construct realistic models.

There is an interesting relation between dimensional reduction of pure Yang-Mills theories and spontaneous compactification in Einstein-Yang-Mills systems. Namely, it turns out that the extrema of the Higgs potential of the reduced Yang-Mills theory correspond to exact solutions of spontaneous compactification type of the coupled Einstein-Yang-Mills system [21], [72-75] . This allows to give a certain physical interpretation to these solutions and is, in particular, helpful in the discussion of their classical stability.

To summarize, the dimensional reduction scheme is an attractive and promising tool for constructing unification models. However, till now nobody succeeded in constructing a model which would be satisfying in all respects. One should also have in mind that there is still a lot of serious open problems related to this method. Here are some of them:

1. The cosmological constant puzzle [6] .
2. The problem of the choice of the vacuum. Usually there is a lot of vacuum solutions of the form E = MxI and there is no

strong criterium which one to choose. One necessary condition for a vacuum to be acceptable is its classical and quantum stability [19,20] .

3. The problem of non-renormalizability of the multidimensional theory and consequences of this fact for the effective theory in four dimensions. Some aspects concerning the contribution of quantum effects have been already studied [6, 76-79].
4. The (partially philosophical) question concerning the dimension D of E. Low energy physics tells us that $D = 4$. A consistent formulation of the theory of superstrings at energies of the order of 10^{19} GeV requires $D = 10$. Thus, maybe the dimension of our real physical world is a function of the energy scale.

Our Review is organized as follows: In chapter 1 we discuss the dimensional reduction method for pure Yang-Mills theories. In section 1.1. we give a classification of G-invariant configurations using the fibre bundle language and in section 1.2. we discuss the reduction of the physical action. For clearness of presentation we restrict ourselves to the case when the metric on $E = M \times G/H$ is of the form $\tilde{\gamma} \oplus \gamma$, with γ being G-invariant. In section 1.3. we discuss the problem of solving the constraint equation for the scalar field. We start with a discussion of some general properties of the scalar field potential, present a general method for calculating the potential, give a full analysis of the structure of this potential for the case of G/H being symmetric and comment, finally, on the irreducibility of scalar field multiplets.

In chapter 2 we deal with dimensional reduction of gravity and spontaneous compactification. In section 2.1. we give a classification of G-invariant configurations for gravity including torsion (Einstein-Cartan-theory), using again fibre bundle methods. In section 2.2. we discuss the dimensional reduction of the Einstein-Hilbert Lagrangian (torsion free case) and in section 2.3. we present a method for finding solutions to the Einstein-Yang-Mills system of spontaneous compactification type, based on a deep relation between dimensional reduction of pure Yang-Mills theories and the spontaneous compactification mechanism.

In chapter 3 we discuss dimensional reduction of matter fields, present the dimensional reduction of the Lagrangian for fermions and comment on some aspects of model building. In particular, we present a few illuminating examples.

We hope that the material contained in this Review is of interest for "pure" physicists as well as for readers who are more interested in the mathematical - or methodological - aspect of the subject. The basic geometrical structures underlying the dimensional reduction method are presented in its whole beauty; on the other hand, for the more pragmatic reader, all important formulae are also given in local terms. We are aware of the fact that some important modern aspects of the dimensional reduction method, like dimensional reduction of supergravity and spontaneous compactification within the theory of superstrings, are not contained in this Review. The main reason for this lack of completeness is that there exist already reviews on these subjects [6,9,32,35,80]. On the other hand, we feel that there is still a lot of open questions in these fields.

1. Dimensional reduction of pure Yang-Mills theories

The most adequate language to describe gauge theories is that of fibre bundles and connections [69,81] . In particular, symmetric gauge fields correspond to invariant connections [39,40] . Here we adopt this point of view and use the fibre bundle approach in the spirit of [69]. Then the first step of the dimensional reduction procedure is a purely geometrical one. It consists in classifying invariant connections and reducing the gauge field action due to the additional space time symmetry. This problem was considered in [42-47]. Here we follow papers [44,46] , which are based on the fibre bundle reduction technique [82] .

1.1. Classification of G-invariant configurations

Let us consider a gauge theory on a principal fibre bundle $\hat{P}(E,K;\pi,\psi)$, where E is the base manifold, K is the structure group (compact connected Lie group), $\pi : \hat{P} \longrightarrow E$ is the canonical projection and $\psi : \hat{P} \times K \longrightarrow \hat{P}$ denotes the free right action of K on $\hat{P}$. The manifold E is orientable and equipped with a (pseudo)-Riemannian metric $\hat{\gamma}$ of signature (-,+,...,+), considered in this chapter as fixed. The group K plays the role of the gauge group, a gauge potential is a connection form on $\hat{P}$ and a matter field is a section in a bundle associated with $\hat{P}$. Sections can be considered as equivariant mappings from $\hat{P}$ to the typical fibre of the associated bundle [69] . Further we shall use this fact.

Let us assume that an additional space time symmetry group G (compact connected Lie group) acts on E to the left,

$$\delta : G \times E \longrightarrow E \quad , \qquad (1.1.1)$$

and let us restrict ourselves to the case, when this action has only one orbit type (H), see [83] . Then E is a fibre bundle over M:= E/G with typical fibre G/H , with H being a represantative of the class (H) . This bundle is associated with the principal bundle W(M,N/H) [83] , where N denotes the normalizer of H in G. In what follows M will play the role of 4-dimensional space time.

To define the notion of a G-invariant connection form on $\hat{P}$, one obviously needs a lift of the G-action to automorphisms of $\hat{P}$,

$$\hat{\delta} : G \times \hat{P} \longrightarrow \hat{P} \quad . \qquad (1.1.2a)$$

Restrictions of $\hat{\delta}$ to the first respectively second component of $G \times \hat{P}$ are denoted by

$$\hat{\delta}_g : \hat{P} \longrightarrow \hat{P} \quad , \qquad (1.1.2b)$$

$$\hat{\delta}_{\hat{p}} : G \longrightarrow \hat{P} \quad . \qquad (1.1.2c)$$

The same notation is used for other group actions. We have $\hat{\delta}_g \in \mathrm{Aut}(\hat{P})$ and $\pi \circ \hat{\delta}_g = \delta_g \circ \pi$, for all $g \in G$.

It is well known that a lift (1.1.2) does not always exist [40], see also [84] . Thus, in a first step one has to sove the following problem: Classify all principal bundles $\hat{P}(E,K)$ with G-action (as automorphisms), projecting onto a given G-action on E. If G acts transitively, the answer is well-known [40,42,82] . Following [85], we give a generalization to the (non-transitive) case, when G acts with one orbit type and the bundle $E \longrightarrow M$ admits a global section

$$s : M \longrightarrow E \quad . \qquad (1.1.3a)$$

(This means that W(M,N/H) is trivial.) Of course, s can be chosen such that

$$\delta(h,s(x)) = s(x) \quad , \qquad (1.1.3b)$$

for all $h \in H$ and $x \in M$. We shall call this case the case of simple action of G on E. In [42] the above classification problem was solved for this case under the additional assumption that M is contractible.

We denote $\tilde{M} := s(M)$ and $\tilde{P} := \pi^{-1}(\tilde{M})$. Obviously, $\tilde{P}(\tilde{M},K)$ is a principal K-bundle over $\tilde{M}$. Restrictions of ψ and π to $\tilde{P}$ - and to other subbundles - will be denoted by the same letters. First suppose that an action $\hat{\delta}$, projecting onto a simple action δ is given and that a section (1.1.3) has been chosen. Observe that the restriction $\tilde{\delta}$ of $\hat{\delta}$ to $H \subset G$ acts on $\tilde{P}$ as a subgroup of $\mathrm{Aut}_0(\tilde{P})$ (vertical automorphisms of $\tilde{P}$), $\tilde{\delta} : H \longrightarrow \mathrm{Aut}_0(\tilde{P})$. We also have the right action of H on the principal bundle $G \longrightarrow G/H$. Thus, we can define

$\chi : G \times_H \tilde{P} \longrightarrow \tilde{P}$, putting

$$\chi([(g,\tilde{p})]) := \tilde{\sigma}(g,\tilde{p}) \quad , \tag{1.1.4}$$

with $(g,\tilde{p})$ being a representative of the class $[(g,\tilde{p})]$. One can easily show that definition (1.1.4) does not depend on the choice of the representative and that χ is a G-equivariant diffeomorphism (with respect to the natural action of G on $G \times_H \tilde{P}$).

Conversely, let there be given a simple G-action on E, a section (1.1.3), a principal K-bundle $\tilde{P}$ and a homomorphism

$$\tilde{\sigma} : H \longrightarrow \mathrm{Aut}_o(\tilde{P}) \quad . \tag{1.1.5}$$

Then one can show that $\hat{P} := G \times_H \tilde{P}$ is naturally a principal K-bundle over E and the natural action of G on $\hat{P}$ projects onto the simple G-action on E.

After this first step we have reduced the lift problem for simple G-action to the problem of analyzing the structure of principal K-bundles $\tilde{P}$ admitting homomorphisms (1.1.5). We solve this problem after making an additional regularity assumption: We define a mapping (associated with $\tilde{\sigma}$),

$\tilde{\tau} : H \times \tilde{P} \longrightarrow K$, putting

$$\tilde{\tau}(h,\tilde{p}) := \psi_{\tilde{p}}^{-1} \circ \tilde{\sigma}(h,\tilde{p}) \quad . \tag{1.1.6a}$$

For every $\tilde{p} \in \tilde{P}$, $\tilde{\tau}_{\tilde{p}} : H \longrightarrow K$ is a group homomorphism and for every h, $\tilde{\tau}_h : \tilde{P} \longrightarrow K$ is an equivariant mapping,

$$\tilde{\tau}_h(\psi_k(\tilde{p})) = k^{-1} \cdot \tilde{\tau}_h(\tilde{p}) \cdot k \quad . \tag{1.1.6b}$$

We denote $\tau := \tilde{\tau}_{\tilde{p}_o}$ for a fixed $\tilde{p}_o \in \tilde{P}$, and assume that $\tilde{\tau}_{\tilde{p}}$ is such that for every $\tilde{p}$ there exists an element $k(\tilde{p})$ of K, with

$$\tilde{\tau}_{\tilde{p}} = k(\tilde{p}) \cdot \tau \cdot k(\tilde{p})^{-1} \quad . \tag{1.1.6c}$$

This assumption means that all the homomorphisms $\tilde{\tau}_{\tilde{p}}$ lie on the same orbit of the action of inner automorphisms in Hom(H,K). Observe that the stabilizer of this action is

$$C \equiv C_K(\tau(H)) := \{ k \in K : k \cdot \tau(h) \cdot k^{-1} = \tau(h) \quad , h \in H \} \quad , \tag{1.1.6d}$$

with C denoting the centralizer of $\tau(H)$ in K. Therefore, we can treat the K-equivariant mapping $\tau : \tilde{P} \longrightarrow \mathrm{Hom}(H,K)$ as a section of the bundle $\tilde{P} \times_K K/C$, associated with $\tilde{P}$. A standard theorem about bundle reductions [82] tells us that $\tilde{P}$ is reducible to a principal C-bundle P, which is given by

$$P := \{\tilde{p} \in \tilde{P}: \quad \tilde{\tau}_{\tilde{p}}(h) = \tau(h) \quad , \quad h \in H\} \quad . \tag{1.1.7a}$$

The mapping $\tilde{\chi} : K \times_C P \longrightarrow \tilde{P}$, defined by

$$\tilde{\chi}([(k,p)]) := \Psi_{k^{-1}}(p) \quad , \tag{1.1.7b}$$

is a K-equivariant diffeomorphism - as one easily shows.(Here K is treated as a principal C-bundle, $K \longrightarrow K/C$.)

Conversely, let there be given a pair (τ,P). Then one can show that $\tilde{P} := K \times_C P$ is naturally a principal K-bundle over $\tilde{M}$ and there exists a natural homomorphism $\tilde{\zeta} : H \longrightarrow \mathrm{Aut}_o(\tilde{P})$, satisfying (1.1.6c).

As a final result we obtain that - for the case of simple G-action and under the additional assumption (1.1.6c) - bundles admitting lifts are classified by pairs (τ,P), where $\tau : H \longrightarrow K$ is a group homomorphism and P is a principal C-bundle over $\tilde{M}$, which can be - of course - nontrivial. For the case of contractible $\tilde{M}$ [42] , (and for a fixed immersion $\tilde{M}$), bundles admitting lifts are simply classified by homomorphisms τ , (as in the case of transitive G-action).

Now, let there be given a bundle $\hat{P}$ admitting a lift of a simple G-action. A connection form $\hat{\omega}$ on $\hat{P}$ is called G-invariant, if

$$\zeta_g^* \omega = \omega \quad . \tag{1.1.8}$$

We shall classify connections fulfilling (1.1.8). For that purpose we choose an AdG-invariant scalar product $\langle \cdot , \cdot \rangle_{\mathfrak{g}}$ on the Lie algebra $\mathfrak{g}$ of G and take the orthogonal with respect to $\langle \cdot , \cdot \rangle_{\mathfrak{g}}$ (reductive) decomposition

$$\mathfrak{g} = \mathfrak{h} \oplus \mathfrak{m} \quad , \tag{1.1.9a}$$

where $\mathfrak{h}$ denotes the Lie algebra of H. This decomposition defines a canonical connection in the bundle $G(G/H,H)$ [82] , and this connection induces a connection in the associated bundle

$G \times_H \tilde{P}$. Taking the image of the corresponding splitting of the tangent bundle $T(G \times_H \tilde{P})$ under $\chi^{`}$, one obtains a canonical splitting of $T\hat{P}$:

$$T_{\hat{p}}\hat{P} = \sigma^{`}_g \{ T_{\tilde{p}}\tilde{P} \oplus \sigma^{`}_{\tilde{p}}(\mathfrak{m}) \qquad , \tag{1.1.9b}$$

where $\hat{p} = \sigma_g(\tilde{p})$ and $\sigma^{`}_{\tilde{p}}(\mathfrak{m})$ is the subspace spanned by Killing fields, generated by elements of $\mathfrak{m}$. We denote the decomposition of 1-forms on $\hat{P}$, corresponding to (1.1.9b), by $\alpha = \alpha^v + \alpha^h$. Obviously, a G-invariant connection form $\hat{\omega}$ is completely given by its values on $\tilde{P}$, and decomposing $\hat{\omega}\restriction_{\tilde{P}}$ with respect to (1.1.9b), we can characterize it by the following objects on $\tilde{P}$:

$$\tilde{\omega} := \omega^v\restriction_{\tilde{P}} \qquad , \tag{1.1.10a}$$

$$\tilde{\Phi}(\tilde{p}) := \sigma^*_{\tilde{p}}\, \omega^h\restriction_{\tilde{P}} \qquad . \tag{1.1.10b}$$

It is easy to show that $\tilde{\omega}$ is a connection form on $\tilde{P}$ [82] , and that $\tilde{\Phi} : \tilde{P} \longrightarrow \mathfrak{m}^* \otimes \mathfrak{k}$ is an equivariant mapping, $\tilde{\Phi} \circ \Psi_k = \mathrm{Ad}k^{-1} \circ \tilde{\Phi}$, ($\mathfrak{k}$ denotes the Lie algebra of K). Moreover, one concludes from (1.1.8) that $(\tilde{\omega}, \tilde{\Phi})$ fulfills:

$$\tilde{\sigma}^*_h \tilde{\omega} = \tilde{\omega} \qquad , \tag{1.1.11a}$$

$$\tilde{\Phi}(\tilde{p}) \circ \mathrm{Ad}h = \mathrm{Ad}\,\tilde{\tau}_{\tilde{p}}(h) \circ \tilde{\Phi}(\tilde{p}) \ , \qquad \text{for all } h \in H. \tag{1.1.11b}$$

If, conversely, a homomorphism $\tilde{\sigma} : H \longrightarrow \mathrm{Aut}_o(\tilde{P})$ and a pair $(\tilde{\omega}, \tilde{\Phi})$ of objects on $\tilde{P}$, fulfilling (1.1.11), are given, then one reconstructs $\hat{P}$ as described above and the G-invariant connection form $\hat{\omega}$ on $\hat{P} = G \times_H \tilde{P}$ as follows: A vector Z, tangent to $G \times_H \tilde{P}$ at $[(g,\tilde{p})]$, is an equivalence class, $Z = [(l^{`}_g(u), Y)]$, with $u \in \mathfrak{g}$, $Y \in T_{\tilde{p}}\tilde{P}$ and l_g denoting left multiplication on G by g. We put

$$\hat{\omega}_{[(g,\tilde{p})]}(Z) := \tilde{\omega}_{\tilde{p}}(Y) + \tilde{\tau}^{`}_{\tilde{p}}(u^{\mathfrak{h}}) + \tilde{\Phi}(\tilde{p})(u^{\mathfrak{m}}), \tag{1.1.12}$$

with $u^{\mathfrak{h}}$ and $u^{\mathfrak{m}}$ denoting the components of u with respect to (1.1.9a). It is easy to verify that $\hat{\omega}$, defined by (1.1.12) has all properties of a connection form and using (1.1.11), one shows that $\hat{\omega}$ is G-invariant.

Thus, our classification problem is reduced to the classification of pairs $(\tilde{\omega},\tilde{\Phi})$ fulfilling (1.1.11). But, from (1.1.6a) and the fact that $\tilde{\omega}$ is a connection form, we have

$$\tilde{\zeta}^*_h \tilde{\omega}\restriction_P = \psi^*_{\tau(h)}\tilde{\omega}\restriction_P = \mathrm{Ad}\,\tau(h^{-1})(\tilde{\omega}\restriction_P) \quad .$$

Using (1.1.11a), we get $\mathrm{Ad}\,\tau(h^{-1})(\tilde{\omega}\restriction_P) = \tilde{\omega}\restriction_P$, for all $h \in H$. This means that $\tilde{\omega}\restriction_P$ takes values in the Lie algebra $\mathfrak{c}$ of C. Therefore, $\tilde{\omega}$ is reducible to P and

$$\omega := \tilde{\omega}\restriction_P \qquad (1.1.13a)$$

is a connection form on P. The restriction

$$\phi := \tilde{\Phi}\restriction_P \qquad (1.1.13b)$$

fulfills - due to (1.1.11b) and (1.1.7a) -

$$\phi(p)\circ \mathrm{Ad}h = \mathrm{Ad}\,\tau(h)\circ\phi(p) \quad . \qquad (1.1.13c)$$

If, conversely, a principal C-bundle P and a pair (ω,ϕ) are given, then one reconstructs $\tilde{P}$, putting $\tilde{P} := K \times_C P$, and from (ω,ϕ), fulfilling (1.1.13c), one easily reconstructs $(\tilde{\omega},\tilde{\phi})$, fulfilling (1.1.11).

Finally, a G-invariant connection form $\hat{\omega}$ on $\hat{P}$ is in 1-1 correspondence with a pair (ω,ϕ) of objects on P, where ω is a connection form and ϕ is an equivariant mapping with values in $\mathfrak{m}^*\otimes\mathfrak{k}$, fulfilling (1.1.13c). Physically, a G-symmetric gauge potential on a multidimensional universe E with values in $\mathfrak{k}$ is completely characterized by a gauge potential on physical space time $\tilde{M}$ with values in the Lie algebra $\mathfrak{c}$ of the reduced gauge group C and a matter field on $\tilde{M}$ with values in $\mathfrak{m}^*\otimes\mathfrak{k}$, fulfilling additional constraint equations (1.1.13c).

We would like to mention that in [45] the classification of G-invariant connections has been given for the case, when W(M,N/H) is nontrivial - but, under the a priori assumption that a lift of the G-action exists. (A classification of bundles admitting lifts, for this case, is not known to us.) In this case a G-invariant connection form splits, of course, also into a pair of quantities - say (ω, ϕ) - but $\hat{P}$ is not reducible to a bundle with structure group C. Therefore, the quantities (ω, ϕ) naturally live on a "bigger" subbundle of $\hat{P}$ and ω cannot be interpreted as a connection form on this bundle. It is interesting to notice that ω can be completed to a connection form on this bundle using the lift of the connection form in W(M,N/H), stemming from the G-invariant metric $\hat{\gamma}$ on E [45].

1.2. Reduction of the physical action

In order to be able to reduce the Yang-Mills action on E, we have to demand that the metric $\hat{\gamma}$ on E is also G-invariant:

$$\delta_g^* \hat{\gamma} = \hat{\gamma} \quad . \tag{1.2.1}$$

If (1.1.8) and (1.2.1) are fulfilled, then we call G a space time symmetry group of the gauge theory under consideration. Thus, the next problem in the dimensional reduction procedure is the classification of pseudo-Riemannian metrics fulfilling (1.2.1). This problem has been solved in [71] for the case of G-action with one orbit type, and without the assumption that a section (1.1.3) exists. The result is the following (compare also with chapter 2): A G-invariant metric $\hat{\gamma}$ is in 1-1 correspondence with a tripel $(\tilde{\gamma}, \xi, \beta)$, where $\tilde{\gamma}$ is the metric induced on M, ξ is an N/H-connection form in W(M,N/H) and β is a matter field on W with values in $\mathfrak{m}^* \overset{s}{\otimes} \mathfrak{m}^*$. If a section (1.1.3) exists, then ξ can be represented by a 1-form on $\tilde{M}$ with values in the Lie algebra $\mathfrak{n}$ of N/H and β by a function on $\tilde{M}$ with values in $\mathfrak{m}^* \overset{s}{\otimes} \mathfrak{m}^*$. In [46] we have performed the reduction of the Yang-Mills action for this case; we shall comment on this at the end of this section. Here we restrict ourselves, for simplicity, to the case, when ξ vanishes. Identifying E with M x G/H via f: M x G/H $\longrightarrow$ E, defined by

$$f(x,[g]) := \delta(g,s(x)) \quad , \tag{1.2.2}$$

one easily calculates

$$(f^* \hat{\gamma})_{(x,[g])} = (s^* \hat{\gamma})_x \oplus (1_{g^{-1}}^* \circ \delta_{s(x)}^* \hat{\gamma})_{[g]} \quad . \tag{1.2.3a}$$

We denote $s^*\hat{\gamma} \equiv \tilde{\gamma}$ and assume, moreover, that $(1^*_{g^{-1}} \circ \delta^*_{s(x)} \hat{\gamma})_{[g]}$ is constant on M. Then we get - writing $\hat{\gamma}$ instead of $f^*\hat{\gamma}$ -

$$\hat{\gamma} = \tilde{\gamma} \oplus \gamma \quad , \qquad (1.2.3b)$$

where γ is a G-invariant metric on G/H. Our further considerations will be for this special form of the metric on E.
We denote by

$$\hat{\Omega} = d\hat{\omega} + 1/2\,[\hat{\omega}, \hat{\omega}] \qquad (1.2.4a)$$

the curvature form of $\hat{\omega}$ and by

$$*\hat{\Omega} = \gamma_{\hat{P}}^{-1}(\hat{\Omega}) \lrcorner\, \vartheta_{\hat{P}} \quad , \qquad (1.2.4b)$$

the dual form in the sense of Hodge. Here $\gamma_{\hat{P}}$ denotes the metric on $\hat{P}$ obtained canonically from $\hat{\gamma}$, $\hat{\omega}$ and the AdK-invariant scalar product $\langle \cdot , \cdot \rangle_{k}$ on k [82], and $\vartheta_{\hat{P}}$ is the volume form corresponding to $\gamma_{\hat{P}}$. (The symbol $\lrcorner$ denotes contraction of multi-vector-fields with differential forms.)
Using that $\hat{\Omega}$ is a horizontal form and that $\vartheta_{\hat{P}}$ naturally decomposes as

$$(\vartheta_{\hat{P}})_{\hat{p}} = (\pi^*\vartheta_E)_{\hat{p}} \wedge (\psi_{\hat{p}}^{-1})^* \vartheta_K \quad , \qquad (1.2.4c)$$

where ϑ_E and ϑ_K are the canonical volume forms on E and K respectively, we get:

$$* \hat{\Omega} = \tilde{*} \hat{\Omega} \wedge (\psi_{\hat{p}}^{-1})^* \vartheta_K \quad , \qquad (1.2.4d)$$

with

$$\tilde{*} \hat{\Omega} = (\pi^*\hat{\gamma})^{-1}(\hat{\Omega}) \lrcorner\, \pi^*\vartheta_E \qquad (1.2.4e)$$

denoting the Hodge dual for horizontal forms.

The gauge field action on the bundle space is given by

$$S = \frac{1}{\mathrm{vol}(K)} \int_{\hat{P}} \hat{\Omega} \dot{\wedge} * \Omega$$

$$= \frac{1}{\mathrm{vol}(K)} \int_{\hat{P}} (\hat{\Omega} \dot{\wedge} \tilde{*} \hat{\Omega})(\psi_{(\cdot)}^{-1})^* \vartheta_K$$

$$= \frac{1}{\mathrm{vol}(K)} \int_{\hat{P}} (\hat{\Omega},\hat{\Omega})\, \vartheta_{\hat{P}} \quad , \tag{1.2.5}$$

where $(\cdot\,,\cdot)_{\hat{p}}$ is the scalar product in the space $\overset{2}{\wedge} H^*_{\hat{p}} \otimes k$, of horizontal 2-forms on $\hat{P}$ with values in k), $\hat{\wedge}$ denotes the exterior product of forms and the scalar product in k and vol(K) is the volume of K. Using (1.1.4) we have

$$S = \frac{1}{\mathrm{vol}(K)} \int_{G x_H \tilde{P}} x^*(\hat{\Omega},\hat{\Omega})\, x^* \vartheta_{\hat{P}} \quad . \tag{1.2.6}$$

It is obvious that

$$(x^* \circ \pi^* \hat{\gamma})_{[(g,\tilde{p})]} = \sigma^*_{g^{-1}} \{ (\pi^* \tilde{\gamma})_{\tilde{p}} \oplus (\bar{\pi}^* \gamma)_{\tilde{p}} \} \quad , \tag{1.2.7a}$$

where $\tilde{p} \equiv [(1_G,\tilde{p})]$ and $\bar{\pi} : G\, x_H\, \tilde{P} \longrightarrow G/H$, and, therefore,

$$(x^* \circ \pi^* \vartheta_M)_{[(g,\tilde{p})]} = \sigma^*_{g^{-1}} \{ (\pi^* \vartheta_M)_{\tilde{p}} \wedge (\bar{\pi}^* \vartheta_{G/H})_{\tilde{p}} \} \quad . \tag{1.2.7b}$$

Moreover, we have

$$(x^* \circ (\psi^{-1}_{\sigma_k(\tilde{p})}) \vartheta_K)_{[(g,\tilde{p})]} = \sigma^*_{g^{-1}} \circ (\psi^{-1}_{\tilde{p}})^* \vartheta_K \quad . \tag{1.2.7c}$$

Using (1.2.7) and the fact that - due to G-invariance of $\hat{\omega}$ - $x^*(\hat{\Omega},\hat{\Omega})$ is constant on G-orbits, we can integrate in (1.2.6) over the orbits. The result is:

$$S = \frac{\mathrm{vol}(G/H)}{\mathrm{vol}(K)} \int_{\tilde{P}} x^*(\hat{\Omega},\hat{\Omega})\, \vartheta_{\tilde{P}} \quad . \tag{1.2.8}$$

Now we decompose $x^*(\hat{\Omega},\hat{\Omega})_{\tilde{p}}$, $\tilde{p} \equiv [(1_G,\tilde{p})]$, due to (1.1.9) and (1.1.10). The result is

$$(x^* \hat{\Omega})^{(v,v)}_{\tilde{p}} = \tilde{\Omega}_{\tilde{p}} \quad , \tag{1.2.9a}$$

$$(x^* \hat{\Omega})^{(v,h)}_{\tilde{p}} = 1/2\, \bar{\pi}^* D\tilde{\Phi}(\tilde{p}) \quad , \tag{1.2.9b}$$

$$(x^* \hat{\Omega})^{(h,h)}_{\tilde{p}} = \bar{\pi}^* \rho(\tilde{\Phi})(\tilde{p}) \quad . \tag{1.2.9c}$$

Here

$$\tilde{\Omega} = d\tilde{\omega} + 1/2 [\tilde{\omega},\tilde{\omega}] \in \overset{2}{\wedge} \tilde{P} \otimes k \tag{1.2.10a}$$

is the curvature form of $\tilde{\omega}$,

$$D\tilde{\Phi} = d\tilde{\Phi} + [\tilde{\omega},\tilde{\Phi}] \in \overset{1}{\Lambda}\tilde{P} \otimes (\mathfrak{m})^* \otimes \mathfrak{k} \quad . \tag{1.2.10b}$$

is the covariant derivative of $\tilde{\Phi}$ with respect to $\tilde{\omega}$, and

$$p(\tilde{\Phi}) = 1/2\{ [\tilde{\Phi},\tilde{\Phi}] - \tilde{\Phi}\circ[\cdot,\cdot]\restriction_{\mathfrak{m}} - \tau^{\cdot}\circ[\cdot,\cdot]\restriction_{\mathfrak{h}}\} \in \mathring{\Lambda}P \otimes \overset{2}{\Lambda}\mathfrak{m}^* \otimes \mathfrak{k} \quad , \tag{1.2.10c}$$

where $[\cdot,\cdot]\restriction_{\mathfrak{m}}$ and $[\cdot,\cdot]\restriction_{\mathfrak{h}}$ denote the restrictions of the commutator in $\mathfrak{g}$ to $\mathfrak{m}$ and $\mathfrak{h}$ respectively.
Inserting (1.2.9) into (1.2.8), we get

$$S = \frac{\mathrm{vol}(G/H)}{\mathrm{vol}(K)} \int_{\tilde{P}} \{ (\tilde{\Omega},\tilde{\Omega}) + 1/2(D\tilde{\Phi},D\tilde{\Phi}) + (p(\tilde{\Phi}),p(\tilde{\Phi}))\} v_{\tilde{P}} \, , \tag{1.2.11}$$

due to (1.2.7a). (For simplicity we have denoted the natural scalar products for vector-space-valued horizontal forms, defined in (1.2.10), all by the same symbol - namely $(\cdot\,,\cdot)$.)

In a final step one reduces S to an action of P, using (1.1.7):

$$S = \frac{\mathrm{vol}(G/H)\mathrm{vol}(K/C)}{\mathrm{vol}(K)} \int_{P} \{(\Omega,\Omega) + 1/2(D\phi,D\phi) + (p(\phi),p(\phi))\} v_P . \tag{1.2.12}$$

As a result of the dimensional reduction procedure we obtain the theory of a Yang-Mills field ω interacting with a matter field ϕ via minimal coupling, defined on reduced space time $\tilde{M}$. The original gauge group K has been reduced to C. The potential term

$$V(\phi) = -(p(\phi),p(\phi)) \tag{1.2.13}$$

is formally of fourth order in ϕ , but ϕ has to fulfill the constraint equation (1.1.13c). Thus, in order to find the explicit form of $V(\phi)$ one has to solve this equation. This is a group-theoretical problem and will be discussed in the next chapter.

In [46] we have performed the reduction of the gauge field action for an arbitrary G-invariant metric $\hat{\gamma}$.

In this case, additional terms in the reduced action appear, describing non-minimal couplings of gauge and matter fields. Since - till now - we were not able to use them for model building, we shall not further comment on them.

Finally, for later purposes, we write down the reduced action in terms of local representatives on reduced space time M. To this end we use the following (local) sections:

$$s_1 \ : \ G/H \longrightarrow G \quad , \tag{1.2.14a}$$

with

$$s_1([\mathbb{1}_G]) = \mathbb{1}_G \quad , \tag{1.2.14b}$$

$$s_1^{\cdot}(T_{[\mathbb{1}G]}\, G/H) = \mathfrak{m} \ , \tag{1.2.14c}$$

$$s_2 \ : \ M \longrightarrow \tilde{P} \quad , \tag{1.2.15a}$$

with

$$s_2(M) \subset P \quad , \tag{1.2.15b}$$

$$\hat{s} \ : \ M \times G/H \longrightarrow \hat{P} \ , \tag{1.2.16a}$$

given by

$$\hat{s}(x,[g]) := \chi([(s_1([g]),s_2(x)])\quad . \tag{1.2.16b}$$

(Here we have again identified E with M x G/H - according to (1.2.2).) Then a simple calculation gives that

$$\hat{A} := \hat{s}^* \hat{\omega} \tag{1.2.17a}$$

is of the following form:

$$\hat{A} = A + \phi(\bar{\Theta}^{\mathfrak{m}}) + \tau(\bar{\Theta}^{\mathfrak{h}})\ , \tag{1.2.17b}$$

where

$$A = s_2^* \omega \quad , \tag{1.2.17c}$$

$$\bar{\Theta} = s_1^*(\Theta)\ , \tag{1.2.17d}$$

with Θ denoting the left-invariant canonical Lie-algebra-valued 1-form on G. The 1-forms $\Theta^{\mathfrak{m}}$ and $\Theta^{\mathfrak{h}}$ are the components of Θ in the sense of (1.1.9a) and the scalar field ϕ stands in

(1.2.17b) and in what follows for the pull-back $s_2^* \phi$. (In other words, the local representative of ϕ will be denoted by the same letter.) For the pull-back

$$\hat{F} := \hat{s}^* \hat{\Omega} \tag{1.2.18a}$$

of the curvature form we obtain

$$\hat{F} = F + (D\phi)(\bar{\Theta}^{\mathfrak{m}}) + 1/2\,[\phi(\bar{\Theta}^{\mathfrak{m}}), \phi(\bar{\Theta}^{\mathfrak{m}})] - 1/2\,\phi([\bar{\Theta}^{\mathfrak{m}}, \bar{\Theta}^{\mathfrak{m}}]_{\restriction \mathfrak{m}})$$
$$- 1/2\,\check{\tau}([\bar{\Theta}^{\mathfrak{m}}, \Theta^{\mathfrak{m}}]_{\restriction \mathfrak{h}}) \quad , \tag{1.2.18b}$$

where

$$F = dA + 1/2\,[A,A] \quad , \tag{1.2.18c}$$

$$(D\phi)(\bar{\Theta}^{\mathfrak{m}}) = (d\phi)(\bar{\Theta}^{\mathfrak{m}}) + [A, \phi(\bar{\Theta}^{\mathfrak{m}})] \quad . \tag{1.2.18d}$$

Choosing a coordinate system $\{x^\mu\}$ on M and a basis $\{u_a\}$ on $\mathfrak{m}$, orthonormal with respect to γ , and denoting $\phi_a \equiv \phi(u_a)$, we obtain the reduced action on M - denoted also by S - as follows:

$$S = \frac{1}{8g^2} \int_M \left\{ - \langle F_{\mu\nu}, F^{\mu\nu} \rangle_{\mathfrak{k}} - 2 \sum_a \langle D_\mu \phi_a, D^\mu \phi_a \rangle_{\mathfrak{k}} - V(\phi) \right\} \vartheta_M \, , \tag{1.2.19a}$$

with

$$V(\phi) = \sum_{a,b} \langle F_{ab}, F^{ab} \rangle_{\mathfrak{k}} \quad , \tag{1.2.19b}$$

$$F_{ab} = [\phi_a, \phi_b] - \phi([u_a, u_b]_{\restriction \mathfrak{m}}) - \check{\tau}([u_a, u_b]_{\restriction \mathfrak{h}}) \, , \tag{1.2.19c}$$

and g denoting the coupling constant of the reduced theory in the "geometrical normalization" (where the coupling constant does not appear in covariant derivatives).

Finally, let us shortly comment on the case $H \cong \{1\!\!1_G\}$, sometimes used in model building [86] . In this case we have, obviously, $C \equiv K$ and $\hat{A} = A + \phi(\Theta)$, with ϕ being in the adjoint representation of $\mathfrak{k}$. If, in particular, $G = U(1) \times \ldots \times U(1)$, then one has on every U(1)-component - in natural angle coordinates - $\Theta = d\alpha$, and, therefore, the coefficients of A do not depend on α .

1.3. Solution of the constraint equation for scalar fields

1.3.1. General properties of the scalar field potential

First we observe that - due to (1.2.19b) we have

$$V(\phi) \geq 0 \quad . \tag{1.3.1a}$$

Defining $\Lambda : M \longrightarrow L(\mathfrak{g}, \mathfrak{k}) = \mathfrak{k} \otimes \mathfrak{g}^*$, by putting

$$\Lambda(u) := \begin{cases} \tau'(u) & \text{for } u \in \mathfrak{h} , \\ \phi(u) & \text{for } u \in \mathfrak{m} , \end{cases} \tag{1.3.1b}$$

we obtain (1.2.19c) in the following form:

$$F_{ab} = [\Lambda(u_a), \Lambda(u_b)] - \Lambda([u_a, u_b]) \quad . \tag{1.3.1c}$$

It follows from (1.2.19b) and (1.3.1c) that $V(\phi) = 0$ if and only if Λ is a homomorphism (considered as a mapping $\Lambda : \mathfrak{g} \longrightarrow \mathfrak{k}$ for every $x \in M$. If Λ is a homomorphism, then the constraint equation (1.1.13c) is automatically fulfilled. Moreover, in this case ϕ corresponds, of course, to the absolute minimum of V and the Lie algebra $\mathcal{R}$ of the group R of gauge transformations, leaving this minimum invariant, is the centralizer of $\Lambda(\mathfrak{g})$ in $\mathfrak{k}$,

$$\mathcal{R} = \{ v \in \mathfrak{k} : \ [v, \Lambda(\mathfrak{g}) = 0 \} \quad . \tag{1.3.2}$$

Since $\tau'(\mathfrak{h}) \subset \Lambda(\mathfrak{g})$, we have $\mathcal{R} \subset \mathcal{L}$. Therefore, in this case the gauge group C of the reduced theory is spontaneously broken to R.

If we want to investigate spontaneous symmetry breaking in the general case, we have to solve equation (1.1.13c) and to find the form of $V(\phi)$ explicitely. For that purpose it is convenient to rewrite the constraint equation in its infinitesimal form:

$$\phi \circ \operatorname{ad} \mathfrak{h} = \operatorname{ad} \tau'(\mathfrak{h}) \circ \phi \quad , \tag{1.3.3a}$$

or, equivalently,

$$\phi([h, u]) = [\tau'(h), \phi(u)] \quad , \ h \in \mathfrak{h}, \ u \in \mathfrak{m} . \tag{1.3.3b}$$

Equation (1.3.3a) has a clear group-theoretical interpretation: ϕ is an operator intertwining the representations $\mathrm{ad}\,\mathfrak{g}_{\restriction \mathfrak{h}}(\mathfrak{m}) \equiv \mathrm{ad}\,\mathfrak{h}(\mathfrak{m})$ and $\mathrm{ad}\,\mathfrak{k}_{\restriction \tau'(\mathfrak{h})}(\mathfrak{k}) \equiv \mathrm{ad}\,\tau'(\mathfrak{h})(\mathfrak{k})$. In order to find ϕ , one has to decompose these representations into irreducible components and then - due to Schur's Lemma - to intertwine equivalent ones. For this purely algebraic discussion , we assume that we have chosen a point $x \in M$, in which $\phi(x): \mathfrak{m} \longrightarrow \mathfrak{k}$ is non-degenerate. For simplicity, we keep on writing ϕ , instead of $\phi(x)$, treating ϕ as an element of $L^H(\mathfrak{m}, \mathfrak{k})$ - the space of matter fields with values in $L(\mathfrak{m}, \mathfrak{k})$ fulfilling (1.3.3).

Since H is compact, we have

$$\mathfrak{h} = \mathfrak{h}' \oplus \mathfrak{z} \quad , \tag{1.3.4a}$$

$$\mathfrak{h}' = \mathfrak{h}_1 \oplus \cdots \oplus \mathfrak{h}_p \quad , \tag{1.3.4b}$$

$$\mathfrak{z} = \mathfrak{z}_1 \oplus \cdots \oplus \mathfrak{z}_q \quad , \tag{1.3.4c}$$

with $\mathfrak{h}_i$ denoting simple (non-Abelian) ideals and $\mathfrak{z}_k$ being 1-dimensional Abelian ideals. Of course, $\mathfrak{h}'$ is the maximal semisimple subalgebra and the $\mathfrak{z}_k$ correspond to the U(1)-factors in H. Irreducible representations of $\mathfrak{h}$ will be characterized by the signature

$$\alpha = [\alpha^{(1)}, \ldots, \alpha^{(p)}]\,(\varkappa^{(1)}, \ldots, \varkappa^{(q)}) \quad ,$$

where $\alpha^{(i)}$ denotes the type of the representation of $\mathfrak{h}_i$ and $\varkappa^{(i)}$ is the eigenvalue of an arbitrary, but fixed element $h_i \in \mathfrak{z}_i$.

Throughout this section we make the following assumptions:

i) $\mathfrak{g}$ and $\mathfrak{k}$ are simple Lie algebras.

ii) The homomorphism $\tau': \mathfrak{h} \longrightarrow \mathfrak{k}$ is injective. (In this case, there always occurs $\mathrm{ad}\,\mathfrak{h}$ in the decomposition of $\mathrm{ad}\,\tau'(\mathfrak{h})(\mathfrak{k})$. The case of τ' not being injective has been discussed in [51]; we shall comment on this in chapter 3.)

iii) In the decompositions of $\mathrm{ad}\,\mathfrak{h}(\mathfrak{m})$ and $\mathrm{ad}\,\mathfrak{h}$ into irreducible components no equivalent representations occur. (For G/H being symmetric or isotropy-irreducible, this assumption is always fulfilled [30] . Moreover, in all interesting examples we have examined, this assumption is also fulfilled.)

Under these assumptions we get

$$\mathrm{ad}\,\mathfrak{h}(\mathfrak{m}) = \zeta \oplus \sum_{k=1}^{N} l_k \alpha_k \quad , \tag{1.3.5a}$$

$$\mathrm{ad}\,\tau(\mathfrak{h})(\mathfrak{k}) = \mathrm{ad}\,\mathfrak{h} \oplus \sum_{j=1}^{p} \mathcal{N}_j \mathrm{ad}\,\mathfrak{h}_j \oplus \sum_{k=1}^{N} n_k \alpha_k \oplus \vartheta \oplus \zeta' . \tag{1.3.5b}$$

Here ζ and ζ' denote trivial representations; due to assumption iii), we have $\dim \zeta = 0$ for $\dim \mathfrak{z} > 0$. By

$$\alpha_k = [\alpha_k^{(1)}, \ldots, \alpha_k^{(p)}] \, (\varkappa_k^{(1)}, \ldots, \varkappa_k^{(q)})$$

we have denoted non-trivial irreducible representations of $\mathfrak{h}$, such that none of them coincides with $\mathrm{ad}\,\mathfrak{h}_i$. The numbers $l_k \geq 1$, $n_k \geq 0$ and $\mathcal{N}_j \geq 1$ denote the multiplicities of these representations and of $\mathrm{ad}\,\mathfrak{h}_j$. The symbol ϑ denotes non-trivial representations, which are not equivalent to α_k and $\mathrm{ad}\,\mathfrak{h}_j$.

The corresponding decompositions of $\mathfrak{m}$ and $\mathfrak{k}$ look as follows:

$$\mathfrak{m} = \mathfrak{n} \oplus \bigoplus_{k=1}^{N} \mathcal{V}^{(k)} \quad , \tag{1.3.6a}$$

$$\mathcal{V}^{(k)} = \bigoplus_{i=1}^{l_k} \mathcal{V}_i^{(k)} \quad , \tag{1.3.6b}$$

$$\mathfrak{k} = \tau(\mathfrak{h}) \oplus \Big(\bigoplus_{j=1}^{p} \mathfrak{U}^{(j)}\Big) \oplus \mathcal{L}' \oplus \Big(\bigoplus_{k=1}^{N} \mathcal{W}^{(k)}\Big) \oplus \mathfrak{k}_{\vartheta} , \tag{1.3.6c}$$

$$\mathfrak{U}^{(j)} = \bigoplus_{t=1}^{\mathcal{N}_j} \mathfrak{U}_t^{(j)} \quad , \tag{1.3.6d}$$

$$\mathcal{W}^{(k)} = \bigoplus_{s=1}^{n_k} \mathcal{W}_s^{(k)} \quad . \tag{1.3.6e}$$

Here $\mathfrak{n}$ denotes the Lie algebra of N/H, the maximal subspace of $\mathfrak{m}$,on which $\mathfrak{h}$ acts trivially. $\mathcal{L}'$ is a subspace carrying trivial representations of $\tau(\mathfrak{h})$ not intersecting with $\tau(\mathfrak{h}) \subset \mathfrak{k}$.

(If $\mathfrak{z} = \{0\}$, we have $\mathcal{L} \equiv \mathcal{L}'$ and for $\mathfrak{z}$ being nontrivial, we get $\mathcal{L} = \mathcal{L}' \oplus \tau'(\mathfrak{z})$. In the latter case we have - in accordance with assumption iii) - $\mathcal{N} = \{0\}$.) The spaces $\mathcal{V}_i^{(k)}$ and $\mathcal{W}_s^{(k)}$ carry the representation α_k, in $\mathcal{U}_t^{(j)}$ we have the representation ad$\mathfrak{h}_j$, and $\mathfrak{k}_\vartheta$ denotes the representation space of ϑ. Without loss of generality we can consider the direct sums in (1.3.6) to be orthogonal in the sense of the Ad-invariant scalar products in $\mathfrak{g}$ and $\mathfrak{k}$.

Now, using Schur's Lemma, we can write down the intertwining operator ϕ in the following form:

$$\phi = \bigoplus_{k=0}^{N} \phi^{(k)} \quad , \tag{1.3.7a}$$

$$\phi^{(o)} : \mathcal{N} \longrightarrow \mathcal{L} \quad , \tag{1.3.7b}$$

$$\phi^{(k)} : \mathcal{V}^{(k)} \longrightarrow \mathcal{W}^{(k)}, \quad k = 1, \dots , N . \tag{1.3.7c}$$

If we denote by

$$E^{(k)s}{}_i : \mathcal{V}_i^{(k)} \longrightarrow \mathcal{W}_s^{(k)} \quad , \tag{1.3.7d}$$

the basic intertwining operators between $\mathcal{V}_i^{(k)}$ and $\mathcal{W}_s^{(k)}$, we get

$$\phi^{(k)} = \sum_{i=1}^{\ell_k} \sum_{s=1}^{n_k} \phi^{(k,i)}{}_s \, E^{(k)s}{}_i \quad . \tag{1.3.7e}$$

(Since trivial representations are 1-dimensional, we have $\ell_o = \dim \mathcal{N}$ and $n_o = \dim \mathcal{L}$ in (1.3.7).) The problem of finding the basic intertwining operators $E^{(k)s}{}_i$ will be studied in the next subsection. The functions $\phi^{(k,i)}{}_s$ make up multiplets of scalar fields with respect to the reduced gauge group C, (every pair (k,i) gives one multiplet). The properties of these multiplets will be discussed in subsection 1.3.3.

In order to make our presentation more transparent, we now make one additional, purely technical assumption:

iv) We assume that $\ell_k = 1$, for all k = 1, ... , N, and $\dim \mathfrak{z} \leq 1$. (The second assumption is reasonable from the physical point of view, because the reduced gauge group C should contain only one U(1)-factor.)

Now, the G-invariant metric γ can be expressed as follows:

$$\gamma_{[1_G]}(u,v) = \sum_{k=0}^{N} m_k^{-2} \langle u^{(k)}, v^{(k)} \rangle_{\mathfrak{g}} \quad , \qquad (1.3.8)$$

with

$$u = \sum_{k=0}^{N} u^{(k)} \quad , \quad v = \sum_{k=0}^{N} v^{(k)} \quad , \quad u^{(o)}, v^{(o)} \in \mathcal{H} \ ,$$

$$u^{(k)}, v^{(k)} \in \mathcal{V}^{(k)} \quad \text{for } k = 1,\dots,N \ .$$

The parameters m_k (with dimension of mass) characterize the geometry of the internal space G/H. Taking an orthonormal (with respect to γ) basis $\{u_a\}$ in $\mathfrak{m}$, denoting $a \equiv (k,r)$ and $u_a \equiv u_{(k,r)}$ with r running, for every fixed k, through the corresponding subspace, we can define

$$v_{(k,r)} := m_k^{-1} u_{(k,r)} \quad . \qquad (1.3.9a)$$

Then $\{v_{(k,r)}\}$ constitute a standard orthonormal basis in $\mathfrak{m}$,

$$\langle v_{(k,r)}, v_{(k',r')} \rangle_{\mathfrak{g}} = \delta_{kk'} \delta_{rr'} \quad . \qquad (1.3.9b)$$

Next we observe that - due to (1.3.3) and the AdK-invariance of $\langle \cdot , \cdot \rangle_{\mathfrak{k}}$ - the mapping

$$(u,v) \longrightarrow \langle \phi(u), \phi(v) \rangle_{\mathfrak{k}} \quad , \qquad (1.3.10a)$$

is an ad$\mathfrak{h}$-invariant scalar product on $\mathfrak{m}$. Denoting the norm in $L^H(\mathfrak{m}, \mathfrak{k})$ by $\|\cdot\|$ and the corresponding norms in subspaces numbered by k, by $\|\cdot\|_k$,

$$\|\phi^{(k)}\|_k^2 := \langle \phi^{(k)}(u^{(k)}), \phi^{(k)}(u^{(k)}) \rangle_{\mathfrak{k}} / \langle u^{(k)}, u^{(k)} \rangle_{\mathfrak{g}} \quad , \qquad (1.3.10b)$$

$$\|\phi\|^2 = \sum_{k=0}^{N} \|\phi^{(k)}\|_k^2 \quad , \qquad (1.3.10c)$$

we get:

$$\langle \phi(u), \phi(v) \rangle_{\mathfrak{k}} = \sum_{k=0}^{N} \|\phi^{(k)}\|_k^2 \langle u^{(k)}, v^{(k)} \rangle_{\mathfrak{g}} \ . \qquad (1.3.10d)$$

Similarly, since $\grave{\tau}$ is a homomorphism and $\langle\,,\,\rangle_{\mathfrak{k}}$ is AdK-invariant, the mapping

$$(h, \grave{h}) \longrightarrow \langle \grave{\tau}(h), \grave{\tau}(\grave{h}) \rangle_{\mathfrak{k}} \quad , \qquad (1.3.11a)$$

$h, \grave{h} \in \mathfrak{h}$, is an ad$\mathfrak{h}$-invariant scalar product on $\mathfrak{h}$. Decomposing h and $\grave{h}$, due to (1.3.4),

$$h = \bigoplus_{i=\varkappa}^{p} h^{(i)} \quad , \quad \grave{h} = \bigoplus_{i=\varkappa}^{p} \grave{h}^{(i)} \quad , \qquad (1.3.11b)$$

with $\varkappa = 1$ for $\mathfrak{z} = \{0\}$ and $\varkappa = 0$ for $\dim \mathfrak{z} = 1$, we have

$$\langle \grave{\tau}(h), \grave{\tau}(\grave{h}) \rangle_{\mathfrak{k}} = \sum_{i=\varkappa}^{p} \lambda_i \langle h^{(i)}, \grave{h}^{(i)} \rangle_{\mathfrak{g}} \quad . \qquad (1.3.11c)$$

(The numbers λ_i characterize the embeddings of $\mathfrak{h}_i$ in $\mathfrak{k}$, they are equal to the ratios of indices of $\mathfrak{h}_i$ in $\mathfrak{k}$ and $\mathfrak{g}$ [87].)

Now we are ready to analyze the general structure of $V(\phi)$, see (1.2.19). For that purpose, it is convenient to expand V in powers of ϕ :

$$\begin{aligned}
V(\phi) &= V^{(0)} + V^{(1)}(\phi) + V^{(2)}(\phi) + V^{(3)}(\phi) + V^{(4)}(\phi) \,, \\
V^{(0)} &= \sum_{a,b} \langle \grave{\tau}([u_a,u_b]\restriction_{\mathfrak{h}}), \grave{\tau}([u_a,u_b]\restriction_{\mathfrak{h}}) \rangle_{\mathfrak{k}} \,, \\
V^{(1)}(\phi) &= 2 \sum_{a,b} \langle \grave{\tau}([u_a,u_b]\restriction_{\mathfrak{h}}), \phi([u_a,u_b]\restriction_{\mathfrak{m}}) \rangle_{\mathfrak{k}} \,, \\
V^{(2)}(\phi) &= \sum_{a,b} \{ \langle \phi([u_a,u_b]\restriction_{\mathfrak{m}}), \phi([u_a,u_b]\restriction_{\mathfrak{m}}) \rangle \\
&\qquad - 2 \langle \grave{\tau}([u_a,u_b]\restriction_{\mathfrak{h}}), [\phi_a, \phi_b] \rangle_{\mathfrak{k}} \} \,, \\
V^{(3)}(\phi) &= -2 \sum_{a,b} \langle \phi([u_a,u_b]\restriction_{\mathfrak{m}}), [\phi_a, \phi_b] \rangle_{\mathfrak{k}} \,, \\
V^{(4)}(\phi) &= \sum_{a,b} \langle [\phi_a, \phi_b], [\phi_a, \phi_b] \rangle_{\mathfrak{k}} \,.
\end{aligned} \qquad (1.3.12)$$

It is easy to calculate $V^{(o)}$ [21,50] :

$$V^{(o)} = \sum_{i=\varkappa}^{p} \lambda_i \sum_{k=1}^{N} d_k m_k^4 C_2^{\mathfrak{h}_i}(\alpha_k^{(i)}) > 0 \quad . \tag{1.3.13}$$

Here $d_k = \dim \alpha_k$ and $C_2^{\mathfrak{h}_i}(\alpha_k^{(i)})$ is the eigenvalue of the quadratic Casimir operator of $\mathfrak{h}_i$, corresponding to the representation $\alpha_k^{(i)}$.

Using the fact that decomposition (1.3.6c) is orthogonal with respect to $\langle \cdot , \cdot \rangle_{\mathfrak{k}}$ and having in mind assumption iii), we conclude that

$$V^{(1)}(\phi) = 0 \quad . \tag{1.3.14}$$

Therefore, $\phi = 0$ is an extremum of $V(\phi)$.

For the quadratic term we obtain after a longer calculation:

$$V^{(2)}(\phi) = \sum_{k=0}^{N} \|\phi^{(k)}\|_k^2 m_k^4 \nu^{(k)} \quad , \tag{1.3.15a}$$

with

$$\nu^{(k)} = \sum_{l,n=0}^{N} m_l^2 m_n^2 m_k^{-4} \sum_{r=1}^{d_l} \sum_{s=1}^{d_n} \sum_{t=1}^{d_k} (c_{(l,r)(n,s)}^{(k,t)})^2$$
$$- 2d_k \sum_{i=\varkappa}^{p} C_2^{\mathfrak{h}_i}(\alpha_k^{(i)})) \quad . \tag{1.3.15b}$$

Here c_{ab}^{d} denote the structure constants of $\mathfrak{g}$ in the basis $\{v_{(k,r)}\}$. Since $V(\phi) \geq 0$, for all ϕ , $\phi = 0$ does not correspond to a minimum of V, if at least one of the $\nu^{(k)}$'s - say $\nu^{(k_o)}$ - fulfills $\nu^{(k_o)} < 0$. In other words, the existence of a number k_o such that $\nu^{(k_o)} < 0$ is a sufficient condition for getting spontaneous symmetry breaking after the reduction to a theory in four dimensions.

Finally, let us comment on two special cases:

a) For the degenerate case $m_k = m$, for all k, one gets:

$$\nu^{(k)} = -4d_k \cdot C_2^{\mathfrak{g}}(ad)(r_k - 1/4) \quad , \qquad (1.3.16a)$$

where $C_2^{\mathfrak{g}}(ad)$ is the eigenvalue of the quadratic Casimir operator of $\mathfrak{g}$ corresponding to the adjoint representation and

$$r_k = \sum_{i=\varkappa}^{p} C_2^{\mathfrak{h}_i}(\alpha_k^{(i)}) \,/\, C_2^{\mathfrak{g}}(ad) \quad . \qquad (1.3.16b)$$

It is easy to show that $0 < r_k \leq 1/2$; and $r_k = 1/2$ if and only if elements of the form $[v_a, v_b]$ do not have a component in $\mathcal{V}^{(k)}$, see [21]. (The necessary information for calculating the quantities $C_2^{\mathfrak{h}_i}(\alpha_k^{(i)})$ and $C_2^{\mathfrak{g}}(ad)$ can be, for example, found in [21,88,89].)
It turns out that for isotropy irreducible internal spaces, one always gets (1.3.16a). In particular, for symmetric spaces one has $r = 1/2$, and the potential of the reduced theory leads to spontaneous symmetry breaking [50,75] ; provided the absolute minimum of $V(\phi)$ is not reached on a scalar field configuration carrying the trivial representation of C.

b) We consider the case, when

$G/H = Sp(2)/(SU(2) \times U(1))$ (6-dimensional non-symmetric

homogeneous space) and $K = SU(5)$. Then $\mathcal{M}$ decomposes into two non-trivial irreducible representations, $N = 2$ and $\mathcal{N} = \{0\}$. If $\tau : SU(2) \times U(1) \longrightarrow SU(5)$ is defined in such a way that one gets $C = SU(3) \times U(1)$, then ϕ intertwins only one of these representations, and we get:

$$V^{(2)}(\phi) = -4 \|\phi\|^2 m_1^4 (2 - \mu) \quad , \qquad (1.3.17)$$

with $\mu = m_2^2 \cdot m_1^{-2}$. That means: If the metric (1.3.8) is such that $\mu < 2$, then the absolute minimum of the potential is reached for $\phi \neq 0$; if $\mu > 2$, then the absolute minimum is obtained for $\phi = 0$.

1.3.2. Method for calculating the potential

In this subsection we present a general method for calculating the scalar field potential. In order to avoid complicated technicalities, we restrict ourselves to the case of G/H being symmetric [30]. We recall that in this case we have $[\mathfrak{m},\mathfrak{m}]\subset\mathfrak{h}$ and - since $\mathfrak{g}$ is simple - $[\mathfrak{m},\mathfrak{m}] = \mathfrak{h}$. Moreover, $\mathfrak{h}$ acts on $\mathfrak{m}$ irreducibly. Consequently, $V^{(3)}(\phi)=0$ in (1.3.12); $V^{(o)}$ and $V^{(2)}(\phi)$ are given by (1.3.13), (1.3.15) and (1.3.16), with $r = 1/2$.

It remains to calculate $V^{(4)}(\phi)$. For that purpose, it is convenient to apply methods of complex simple Lie algebras [91,92] . (In that case we have, for example, always the same intertwining number, namely 1 [90].) Thus, we complexify $\mathfrak{g}$ and $\mathfrak{k}$ and continue ϕ and the constraint equation (1.3.3) linearly to the complexified spaces $\mathfrak{h}^{\mathbb{C}}$, $\mathfrak{m}^{\mathbb{C}}$ and $\mathfrak{k}^{\mathbb{C}}$ [48-50] . The complexified operator has the following reality property:

$$\overline{\phi^{\mathbb{C}}(u)} = \phi^{\mathbb{C}}(\bar{u}) \quad , \quad u\in\mathfrak{m}^{\mathbb{C}} \quad , \tag{1.3.18a}$$

with bar denoting complex conjugation. Obviously, we find

$$\phi = \phi^{\mathbb{C}}\restriction_{\mathfrak{m}} \, , \, \mathfrak{m}\subset\mathfrak{m}^{\mathbb{C}} \quad . \tag{1.3.18b}$$

In the remaining part of this subsection we will work only with complex algebras and, therefore, we omit the sign "C" .

If G/H is symmetric, then $\mathfrak{h}$ is the sum of no more than three ideals [30] :

$$\mathfrak{h} = \mathfrak{h}_o \oplus \mathfrak{h}_1 \oplus \mathfrak{h}_2 \quad , \tag{1.3.19}$$

where $\mathfrak{h}_o$ denotes a 1-dimensional center, and $\mathfrak{h}_1, \mathfrak{h}_2$ are simple subalgebras. If $\mathfrak{h}$ is semisimple, then $\mathfrak{h}$ acts on $\mathfrak{m}\equiv\mathfrak{f}=\overline{\mathfrak{f}}$ irreducibly; if $\mathfrak{h}$ contains a 1-dimensional center $\mathfrak{h}_o$, we have $\mathfrak{m} = \mathfrak{f}\oplus\overline{\mathfrak{f}}$, with $\mathfrak{f}$ being an irreducible subspace. In the latter case we have [92] :

$$\dim\mathfrak{f} = 1/2 \dim\mathfrak{m} \, ,$$

$$[\mathfrak{f},\mathfrak{f}] = 0 \quad , [\mathfrak{f},\overline{\mathfrak{f}}] = \mathfrak{h} \quad . \tag{1.3.20}$$

Let us denote by

$$\mathfrak{f}\otimes\bar{\mathfrak{f}} \longrightarrow \sigma(\mathfrak{f}\otimes\bar{\mathfrak{f}}) \qquad (1.3.21)$$

an operation on $\mathfrak{f}\otimes\bar{\mathfrak{f}}$,which for the case $\mathfrak{f}=\bar{\mathfrak{f}}$ coincides with the antisymmetrization operator and which for $\mathfrak{f}\neq\bar{\mathfrak{f}}$ means the identity. Now we define the following linear mappings:

$f_1 : \sigma(\mathfrak{f}\otimes\bar{\mathfrak{f}}) \longrightarrow \mathfrak{h}$, given by

$$f_1(\sigma(u\otimes v)) := [u,v] \quad , \qquad (1.3.22a)$$

$f_2 : \sigma(\mathfrak{f}\otimes\bar{\mathfrak{f}}) \longrightarrow \mathfrak{k}$, given by

$$f_2(\sigma(u\otimes\bar{v})) := [\phi(u),\phi(\bar{v})] \quad , \quad u,v\in\mathfrak{f} . \qquad (1.3.22b)$$

Since $[\mathfrak{m},\mathfrak{m}]=\mathfrak{h}$, f_1 is surjective. Thus, we have

$$\sigma(\mathfrak{f}\otimes\bar{\mathfrak{f}}) = \mathrm{ker}f_1 \oplus \{\sigma(\mathfrak{f}\otimes\bar{\mathfrak{f}})/\mathrm{ker}f_1\} \quad , \qquad (1.3.22c)$$

and we can define the isomorphism $\hat{f}_1 : \sigma(\mathfrak{f}\otimes\bar{\mathfrak{f}})/\mathrm{ker}f_1 \longrightarrow \mathfrak{h}$,

$$\hat{f}_1 := f_1\restriction \sigma(\mathfrak{f}\otimes\bar{\mathfrak{f}})/\mathrm{ker}f_1 \quad . \qquad (1.3.22d)$$

Similarly, we can define $\hat{f}_2 : \sigma(\mathfrak{f}\otimes\bar{\mathfrak{f}})/\mathrm{ker}f_2 \longrightarrow \mathfrak{k}$,

$$\hat{f}_2 := f_2\restriction \sigma(\mathfrak{f}\otimes\bar{\mathfrak{f}})/\mathrm{ker}f_2 \quad . \qquad (1.3.22e)$$

Now, assuming that

$$\mathrm{ker}f_1 \subset \mathrm{ker}f_2 \quad , \qquad (1.3.22f)$$

and, therefore, $\sigma(\mathfrak{f}\otimes\bar{\mathfrak{f}})/\mathrm{ker}f_2 \subset \sigma(\mathfrak{f}\otimes\bar{\mathfrak{f}})/\mathrm{ker}f_1$, we finally obtain the linear mapping f :   , putting

$$f := \hat{f}_2\circ\hat{f}_1^{-1} \quad . \qquad (1.3.22g)$$

It follows from (1.3.3) that f is an operator intertwining the representations ad $\mathfrak{h}$ and ad $\tau(\mathfrak{h})(\mathfrak{k})$. (Assumption (1.3.22f) turns out to be not very restrictive; it holds in all examples of model building known to us.)

Let us denote

$$\mathfrak{U}^{(0)} := \mathcal{L}' \quad . \qquad (1.3.23a)$$

Then, using (1.3.6) and (1.3.19), we obtain from Schur's Lemma:

$$f : \mathfrak{h} \longrightarrow \bigoplus_{i=0}^{2} (\tau'(\mathfrak{h}_i) \oplus \mathcal{U}^{(i)}) \quad , \qquad (1.3.23b)$$

$$f = \bigoplus (\nu_i \tau'_i + f_i) \quad , \qquad (1.3.23c)$$

where $\tau'_i := \tau'|_{\mathfrak{h}_i}$ and f_i denotes the $\mathcal{U}^{(i)}$-component of f. For the coefficients ν_i one can show [50] :

$$\lambda_o \nu_o = \lambda_1 \nu_1 = \lambda_2 \nu_2 = \|\phi\|^2 \quad , \qquad (1.3.23d)$$

with λ's defined in (1.3.11c).

We define the norm of f_i, putting

$$\| f_i \|^2 = \langle f_i(h), f_i(h') \rangle_{\mathfrak{k}} / \langle h, h' \rangle_{\mathfrak{g}} \quad , \qquad (1.3.24a)$$

with $h, h' \in \mathfrak{h}_i$, $\langle h, h' \rangle_{\mathfrak{g}} \neq 0$. Obviously, $\|f_i\|^2$ is an invariant of fourth order in ϕ , which can be decomposed as follows:

$$\| f_i \|^2 = a_i \|\phi\|^4 + I_i(\phi) \quad . \qquad (1.3.24b)$$

(The real numbers a_i and the invariant $I_i(\phi)$, different from $\|\phi\|^4$, have to be calculated for every example separately.)

Now, let us concentrate on the case $\mathfrak{F} \neq \overline{\mathfrak{F}}$. For symmetric spaces we always have N = 1 in (1.3.6). Moreover, for the case under consideration, we have to replace in these formulae $\mathfrak{F}$ by $\mathfrak{F} \oplus \overline{\mathfrak{F}}$ and $\mathcal{W}_s$ by $\mathcal{W}_s \oplus \overline{\mathcal{W}}_s$. Then we get

$$\phi : \mathfrak{F} \oplus \overline{\mathfrak{F}} \longrightarrow \bigoplus_{s=1}^{n} (\mathcal{W}_s \oplus \overline{\mathcal{W}}_s) \quad . \qquad (1.3.25a)$$

Introducing basic intertwining operators $E^s : \mathfrak{F} \longrightarrow \mathcal{W}_s$ and $\bar{E}^s : \mathfrak{F} \longrightarrow \overline{\mathcal{W}}_s$, normalized by $\langle E^s(u), \bar{E}^t(v) \rangle_{\mathfrak{k}} = \delta^{st} \langle u, v \rangle_{\mathfrak{g}}$, we have

$$\phi = \bigoplus_{s=1}^{n} (\phi_s E^s \oplus \overline{\phi}_s \bar{E}^s) \quad . \qquad (1.3.25b)$$

Of course,

$$\|\phi\|^2 = \sum_{s=1}^{n} \|\phi_s\|^2 \quad , \tag{1.3.25c}$$

which is a special case of (1.3.10c).

Inserting the expression for ϕ given by (1.3.25b) into (1.3.12) and using (1.3.23) and (1.3.24), we obtain [50] :

$$V(\phi) = m^4 \left\{ c_1 (\|\phi\|^2 - c_2/c_1)^2 + I(\phi) + c_o \right\} \quad , \tag{1.3.26a}$$

where

$$I(\phi) = \tilde{d} \sum_{i=0}^{2} C_2^{\mathfrak{h}_i}(\alpha^{(i)})(2 - \delta_{oi}\, d/2)\, I_i(\phi) \quad , \tag{1.3.26b}$$

with $\tilde{d} = d$ in the case under consideration.
The real numbers c_o, c_1 and c_2 are given by :

$$c_o = d \sum_{i=0}^{2} \lambda_i\, C_2^{\mathfrak{h}_i}(\alpha^{(i)}) - c_2^2/c_1 \quad , \tag{1.3.26c}$$

$$c_1 = \tilde{d} \sum_{i=0}^{2} C_2^{\mathfrak{h}_i}(\alpha^{(i)})(1 + a_i \lambda_i)\lambda_i^{-1}(2 - \delta_{oi}\, d/2) \; , \tag{1.3.26d}$$

$$c_2 = d/2 \cdot C_2^{\mathfrak{g}}(ad) \quad . \tag{1.3.26e}$$

It turns out that these formulae are also valid for the simpler case $\mathfrak{F} = \bar{\mathfrak{F}}$, provided we set $\tilde{d} = d/2$.

We end up with a short discussion of two important special cases:

a) Let $\mathfrak{h}$ be simple and $\mathfrak{k}$ contain no other subspaces, except $\tau'(\mathfrak{h})$, carrying the adjoint representation of $\mathfrak{h}$. Then we get:

$$[\phi(u), \phi(v)] = \lambda^{-1}\|\phi\|^2 \cdot \tau'([u,v]) \quad , \tag{1.3.27a}$$

$u, v \in \mathfrak{M} = \mathfrak{F} = \bar{\mathfrak{F}}$. Therefore, we simply have:

$$V(\phi) = d\cdot\lambda^{-1} m^4\, C_2^{\mathfrak{h}}(\alpha)(\|\phi\|^2 - \lambda)^2 \quad . \qquad (1.3.27b)$$

We see that $V(\phi)$ is a canonical Higgs potential, which attains its minimum for $\|\phi\|^2 = \lambda$. From (1.3.27a) we see that on the subspace defined by this equation, ϕ extends $\tau' : \mathfrak{h} \longrightarrow \mathfrak{k}$ to a homomorphism $\Lambda : \mathfrak{g} \longrightarrow \mathfrak{k}$ - in accordance with the general discussion in subsection 1.3.1.

b) Let $\mathfrak{h} = \mathfrak{h}_0 \oplus \mathfrak{h}_1$ and $\mathfrak{k}$ contain no subspaces, except $\tau(\mathfrak{h}_1)$, carrying the adjoint representation of $\mathfrak{h}_1$ and the trivial representation of $\mathfrak{h}_0$. Then

$$[\phi(u), \phi(\bar{v})] = \nu_0 \tau'_0([u,\bar{v}]\restriction_{\mathfrak{h}_0}) + \nu_1 \tau'_1([u,v]\restriction_{\mathfrak{h}_1}) + f_0([u,v]\restriction_{\mathfrak{h}_0}) \quad , \qquad (1.3.28a)$$

$u \in \mathfrak{F}$, $v \in \overline{\mathfrak{F}}$, with f_0 being the restriction of f to $\mathfrak{h}_0$, $f_0 : \mathfrak{h}_0 \longrightarrow \mathfrak{L}$. If we additionally assume that $[\phi(\mathfrak{F}), \phi(\mathfrak{F})] = 0$, then we get:

$$V(\phi) = d\, m^4 C_2^{\mathfrak{h}_0}(\alpha^{(0)})(1/2\cdot\lambda_1^{-1}(\|\phi\|^2 - \lambda_1)^2 + \lambda_0 - \lambda_1) \; . \qquad (1.3.28b)$$

The method of calculating the scalar field potential,developed in this subsection for symmetric spaces, can be - in principle - generalized to arbitrary homogeneous spaces. To this end one has to define mappings (1.3.22) for each pair of irreducible representations in the decompositions (1.3.6a). Moreover, one has to analyze the existence of a mapping f: $\mathfrak{g} \longrightarrow \mathfrak{k}$, analogous to (1.3.22g), to find its general form and to substitute it into (1.3.12). Some special cases of G/H being non-symmetric were discussed in [51].

Let us, finally, note that the problem of decomposing $\mathfrak{g}$ and $\mathfrak{k}$ into irreducible subspces of $\mathfrak{h}$ and $\tau'(\mathfrak{h})$, in particular, that of finding the centralizer of $\tau'(\mathfrak{h})$, as well as the explicit construction of ϕ can be effectively solved with the use of root techniques for complex Lie algebras [87,91,92] , including the technique of root lattices [48,49,51].

1.3.3. Irreducibility of scalar field multiplets

Let $\phi \in L^H(\mathcal{M},\mathcal{K})$ and let $\phi' = \operatorname{ad}\lambda(\phi) = [\lambda,\phi]$, $\lambda \in \mathcal{L}$. It follows directly from the definition of the centralizer that ϕ' also belongs to $L^H(\mathcal{M},\mathcal{K})$. This means that $\operatorname{ad}\mathcal{K}|_{\mathcal{L}}$ acts on $L^H(\mathcal{M},\mathcal{K})$; and if $\phi(u)$ belongs to an invariant subspace $\mathcal{W}^{(k)}$ of $\operatorname{ad}\tau'(\eta)(\mathcal{K})$, then also $\phi'(u)$ belongs to this subspace. In other words, in the notation (1.3.7),

$$\phi_s^{(k,i)} \quad , \; s = 1, \ldots, n_k \, ,$$

are matter field multiplets with respect to $\mathcal{L}$ for every fixed (k,i). However, in general, these multiplets are reducible. In this subsection we distinquish a certain class of multidimensional theories, for which we have only one irreducible multiplet.

It follows from the general theory [87] that a semisimple subalgebra $\mathcal{R}$ of a simple Lie algebra $\mathcal{G}$ is fixed - up to equivalence - by the decomposition of one of the (irreducible) defining representations $\mathcal{D}_{\mathcal{G}}$ of $\mathcal{G}$ into irreducible components with respect to $\mathcal{R}$. We restrict ourselves to the classical series A_n, B_n, C_n and D_n, with the corresponding compact forms SU(n+1), SO(2n+1), Sp(n) and SO(2n). In this case, the defining representation is the lowest-dimensional non-trivial representation; that means $\mathcal{D}_{\mathcal{G}}$ is the fundamental representation for A_n ($\dim \mathcal{D}_{\mathcal{G}} = n+1$) and the vector representation for B_n ($\dim \mathcal{D}_{\mathcal{G}} = 2n+1$), C_n ($\dim \mathcal{D}_{\mathcal{G}} = 2n$) and D_n ($\dim \mathcal{D}_{\mathcal{G}} = 2n$). The adjoint representation of $\mathcal{G}$ can be expressed in terms of $\mathcal{D}_{\mathcal{G}}$ as follows:

$$\operatorname{ad}A_n = \mathcal{D}_{A_n} \,\tilde{\otimes}\, \mathcal{D}^*_{A_n} \quad , \tag{1.3.29a}$$

$$\operatorname{ad}B_n(D_n) = \mathcal{D}_{B_n(D_n)} \overset{a}{\otimes} \mathcal{D}_{B_n(D_n)} \quad , \tag{1.3.29b}$$

$$\operatorname{ad}C_n = \mathcal{D}_{C_n} \overset{s}{\otimes} \mathcal{D}_{C_n} \quad , \tag{1.3.29c}$$

where * denotes the contragradient representation, "~" means removing the 1-dimensional (trivial) representation and "a" and "s" denote the antisymmetrization respectively symmetrization operations in the tensor product.

The decomposition of the defining representation $\mathcal{D}_{\mathcal{G}}$ allows a classification of subalgebras $\mathcal{R} \subset \mathcal{G}$ according to the number of non-trivial irreducible representations of every simple ideal of $\mathcal{R}$ in this decomposition. Using (1.3.29) one can, in a next step, investigate $ad\,\mathcal{G}\restriction_{\mathcal{R}}(\mathcal{G})$, which is relevant for the discussion of matter field multiplets. To demonstrate how the method works, let us consider the case of $\mathcal{R}$ being simple. We have for $\mathcal{G} = A_n$

$$\mathcal{D}_{\mathcal{G}}\restriction_{\mathcal{R}} = \alpha_1 \oplus \ldots \oplus \alpha_k \oplus \zeta \quad , \tag{1.3.30a}$$

and for $\mathcal{G} = B_n, C_n$ or D_n :

$$\mathcal{D}_{\mathcal{G}}\restriction_{\mathcal{R}} = \{\alpha_1\} \oplus \ldots \oplus \{\alpha_k\} \oplus \zeta \quad , \tag{1.3.30b}$$

where $\{\alpha_i\} = \alpha_i$ for $\alpha_i = \alpha_i^*$ and $\{\alpha_i\} = \alpha_i + \alpha_i^*$ for $\alpha_i \neq \alpha_i^*$. Here α_i denote non-trivial, irreducible representations of $\mathcal{R}$ and ζ denotes a trivial one. One calls subalgebras giving (1.3.30), k-subalgebras and the corresponding embeddings k-embeddings.

Let us restrict ourselves further on to the case when $\mathcal{G} = A_n$ and k=1 in (1.3.30). Then we obtain from (1.3.29) and (1.3.30):

$$\begin{aligned} adA_n\restriction\mathcal{R} &= (\mathcal{D}_{A_n} \tilde{\otimes} \mathcal{D}^*_{A_n})\restriction\mathcal{R} \\ &= (\alpha \tilde{\otimes} \alpha^*) \oplus (\alpha \otimes \zeta^*) \oplus (\zeta \otimes \alpha^*) \oplus (\zeta \otimes \zeta^*) + \mathcal{X} \ . \end{aligned} \tag{1.3.31a}$$

It is easy to see that $\zeta \tilde{\otimes} \zeta^*$ carries a trivial representation of $\mathcal{R}$ and belongs to the centralizer $\mathcal{Z}$ of $\mathcal{R}$ in $\mathcal{G}$; $\mathcal{X}$ is another trivial representation of dimension 1. The representation $\alpha \tilde{\otimes} \alpha^*$ contains the adjoint representation of $\mathcal{R}$ [93], and, maybe, also further non-trivial representations of $\mathcal{R}$. The decomposition of $\mathcal{G}$, corresponding to (1.3.31a), is given by:

$$\mathcal{G} = (\mathcal{W}_\alpha \tilde{\otimes} \overline{\mathcal{W}}_\alpha) \oplus (\mathcal{W}_\alpha \otimes \overline{\mathcal{W}}_\zeta) \oplus (\mathcal{W}_\zeta \otimes \mathcal{W}_\alpha) \oplus (\mathcal{W}_\zeta \otimes \mathcal{W}_\zeta) \oplus \mathcal{W}_{\mathcal{X}} , \tag{1.3.31b}$$

where $\mathcal{W}_\alpha, \mathcal{W}_\zeta$ and $\mathcal{W}_\chi$ are the representation spaces of α, ζ and χ. One can show that $\mathcal{L} = \mathcal{L}' \oplus \mathcal{W}_\chi$, $\mathcal{L}' = \mathcal{W}_\zeta \tilde{\otimes} \overline{\mathcal{W}}_\zeta$. In our case, we have $\mathcal{L}' = A_{l-1}$, $l = \dim \mathcal{W}_\zeta$. It is also easy to convince oneself that the spaces $\mathcal{W}_\alpha \otimes \overline{\mathcal{W}}_\zeta$ and $\mathcal{W}_\zeta \otimes \overline{\mathcal{W}}_\alpha$ decompose into sums of subspaces $\mathcal{W}_s$ and $\overline{\mathcal{W}}_s$, $s = 1, \ldots, \dim \mathcal{W}_\zeta$, each carrying the irreducible representation α. The centralizer $\mathcal{L}$ acts transitively on $\{\mathcal{W}_s\}$ and $\{\overline{\mathcal{W}}_s\}$.

Since we are interested in the classification of multiplets under $\mathcal{L}$, it is convenient to restrict $\mathrm{ad}A_n$ to $\mathcal{K} \oplus \mathcal{L}$. One gets from (1.3.31a) :

$$\mathrm{ad}A_n\restriction_{\mathcal{K}\oplus\mathcal{L}} = [\alpha\tilde{\otimes}\alpha^*, 0](0) \oplus [\alpha^*, \beta](1) \oplus [\alpha, \beta^*](-1) \oplus \mathrm{ad}\,\mathcal{L} \quad , \tag{1.3.32}$$

where $\beta \equiv \mathcal{D}_{A_{l-1}}$ is the irreducible representation of $\mathcal{L}'$ of dimension $l = \dim \mathcal{W}_\zeta$. (In this formula again the convention for characterizing the type of representations as explained in subsection 1.3.1 has been used.)

Now, let us investigate the question of irreducibility of ϕ for the case G/H being symmetric. At the beginning, let $\mathfrak{h}$ be simple. Then $\alpha \equiv \mathrm{ad}\,\mathfrak{h}(\mathfrak{m})$ is irreducible and different from the trivial representation. Moreover, $\alpha \tilde{\otimes} \alpha^*$ contains only one representation of type $\mathrm{ad}\,\mathfrak{h}$, (that means $N_j = 0$ in (1.3.6c) and $f_i = 0$ in (1.3.23c)), and does not contain α [30]. Let $\tau'(\mathfrak{h}) \subset \mathfrak{k}$ be a 1-embedding. Then decomposition (1.3.30) is given by

$$\mathcal{D}_{\mathfrak{k}}\restriction_{\tau'(\mathfrak{h})} = \alpha' \oplus \zeta' \quad . \tag{1.3.33}$$

We have to consider two cases:

a) α is equivalent to α'. Then ϕ maps from $\mathfrak{m}$ to $(\mathcal{W}_{\alpha'} \otimes \overline{\mathcal{W}}_{\zeta'}) \oplus (\mathcal{W}_{\zeta'} \otimes \overline{\mathcal{W}}_{\alpha'})$ and the space $L^H(\mathfrak{m}, \mathfrak{k})$ carries the irreducible representation $[\beta](1)$ of $\mathcal{L}$.

b) α and α' are not equivalent. Then ϕ can intertwine only $\mathrm{ad}\,\mathfrak{h}(\mathfrak{m})$ and $\alpha' \tilde{\otimes} (\alpha')^*$. But from (1.3.32) we see that $\alpha' \otimes (\alpha')^*$ belongs to the trivial representation of $\mathcal{L}$, i.e. in this case the space $L^H(\mathfrak{m}, \mathfrak{k})$ carries the trivial representation of $\mathcal{L}$.

In an analogous way [50] one can discuss the case of 1-subalgebras for the general form of $\mathfrak{h}$, see (1.3.19). Also one can do the same analysis for the remaining classical groups, $\mathcal{G} = B_n$, C_n and D_n [50]. As a result of this discussion one gets: In the case of G/H being symmetric and K being a classical Lie group, a sufficient condition to obtain one irreducible (non-trivial) multiplet of scalar fields is that $\tau'(\mathfrak{h})$ is a 1-subalgebra of $\mathfrak{k}$ and α' in (1.3.33) is equivalent to $\alpha \equiv ad_{\mathfrak{h}}(\mathfrak{m})$. (It was shown in [48] that in the case of regular embeddings of $\tau'(\mathfrak{h})$ in $\mathfrak{k}$, this condition is always fulfilled.)

Concrete examples of models containing one irreducible multiplet will be discussed in chapter 3.

Finally, we would like to mention that the method described here can be extended to exceptional Lie algebras $\mathfrak{k}$. In this case, however, one does not have simple formulae of the type of (1.3.29) connecting the adjoint and defining representations [93] . Therefore, in this case, every model has to be studied separately.

2. Dimensional reduction of gravity and spontaneous compactification

In this chapter we discuss the dimensional reduction of gravity, including - at least in the first part - torsion, (Einstein-Cartan theory [94]). Theories with torsion on a multidimensional universe were discussed in the literature in connection with the problem of getting massless chiral fermions [95] and in the context of spontaneous compactification [96,97]. It turns out that including torsion, one can get zero modes for the mass operator (the Dirac operator on the internal space), but for the solution of the chirality problem torsion usually does not help - due to index theorem arguments. (Only when the torsion components on the internal space belong to a topologically non-trivial metric connection, there can be some hope to obtain chiral fermions [96].) All attempts to get torsion-induced spontaneous compactification are in our opinion on a rather heuristic level [97], and need further investigations. (For a first, more systematic step in this direction see [98].)

We conclude that - from the physical point of view - dimensional reduction of Einstein-Cartan-theory may be of interest and, therefore, it seems to be worthwile to study such theories more systematically.

2.1. Classification of G-invariant configurations of Einstein-Cartan theory

Our treatment is based on [99]. In order to avoid certain technicalities (and only for that reason) we make the following simplifying assumptions:

i) Let G act on E to the left, $\delta : G \times E \longrightarrow E$, without fixed points.

ii) Let there exist a section in the bundle $E \longrightarrow M = E/G$.

Then E has the structure of a trivial principal G-bundle over M, with right action $\tilde{\delta}$ of G, defined by $\tilde{\delta}_g := \delta_{g^{-1}}$.

A configuration of Einstein-Cartan theory on E is a pair $(\hat{\omega}, \hat{\gamma})$, where $\hat{\gamma}$ is a (pseudo)-Riemannian metric an E and $\hat{\omega}$ is a linear connection on E, compatible with $\hat{\gamma}$,

$$D\hat{\gamma} = 0 \quad . \qquad (2.1.1a)$$

We shall treat $\hat{\omega}$ as a connection form in L(E) and $\hat{\gamma}$ as a GL(n,R)-equivariant mapping

$$\hat{\gamma} \; : \; L(E) \longrightarrow (R^D)^* \overset{s}{\otimes} (R^D)^* \quad . \qquad (2.1.1b)$$

(Then $D\hat{\gamma}$ is the ordinary covariant derivative of the matter field $\hat{\gamma}$ with respect to the gauge potential $\hat{\omega}$.)

Denoting the natural lift of δ to L(E) by

$$\hat{\delta} \; : G \times L(E) \longrightarrow L(E) \; , \qquad (2.1.2)$$

we can define the notion of a G-invariant configuration of Einstein-Cartan-theory:

$$\hat{\delta}_g^* \, \hat{\omega} = \hat{\omega} \quad , \qquad (2.1.3a)$$

$$\hat{\delta}_g^* \, \hat{\gamma} = \hat{\gamma} \quad . \qquad (2.1.3b)$$

Now we shall classify pairs $(\hat{\omega}, \hat{\gamma})$ satisfying (2.1.1a) and (2.1.3).

First one shows that $\hat{\gamma}$ and the action of G induce the following sequence of bundle reductions:

$$L(E) \longrightarrow \hat{O}(E) \longrightarrow \tilde{O}(E) \longrightarrow O(E) \quad , \qquad (2.1.4)$$

where $\hat{O}(E)$ is the bundle of orthonormal (with respect to $\hat{\gamma}$) frames on E, $\tilde{O}(E)$ is the bundle of adapted orthonormal frames, (principal bundle with structure group O(1,3) x O(d)), and O(E) is the bundle of adapted orthonormal frames with fixed component in the direction of the internal space, (principal bundle over E with structure group O(1,3)). The first reduction is standard:

$$\hat{O}(E) := \left\{ e \in L(E) : \quad \hat{\gamma}(e) = \eta \right\} \quad , \qquad (2.1.5a)$$

where $\eta \in (R^D)^* \overset{s}{\otimes} (R^D)^*$ is in the standard basis of R^D given by $\eta = \mathrm{diag}(-1,+1,\ldots,+1)$. The second reduction is induced by the following splitting of the tangent bundle:

$$TE = \mathcal{V} \oplus \mathcal{H} \quad , \qquad (2.1.5b)$$

with

$$\mathcal{V}_y := \tilde{\delta}'_y(\mathfrak{g}) \quad , \quad y \in E \quad , \tag{2.1.5c}$$

and $\mathcal{H}$ being the orthogonal (with respect to $\hat{\gamma}$) complement of $\mathcal{V}$. Splitting (2.1.5b) is a section of the associated bundle $\hat{O}(E) \times_{O(1,D-1)} G_{4,D}$, with $G_{4,D} := O(1,D-1)/(O(1,3)\times O(d))$ being the space of orthogonal with respect to η decompositions of R^D. Treating this section as an equivariant mapping $\chi : \hat{O}(E) \longrightarrow G_{4,D}$, and fixing one decomposition χ_o , given by

$$R^D = R^4 \oplus R^d \quad , \tag{2.1.5d}$$

we put

$$\tilde{O}(E) := \{ e \in \hat{O}(E) : \quad \chi(e) = \chi_o \} \quad . \tag{2.1.5e}$$

Due to assumption ii), one can perform a standard orthonormalization procedure for the Killing fields of G on a chosen section in $E \longrightarrow E/G$, and in a next step, one can transport them along the fibres by the group action. The result is a section in the bundle of vertical orthonormal frames, which induces a reduction of $\tilde{O}(E)$ to the G-invariant subbundle O(E).

Obviously, O(E) is a principal bundle with structure group G over O(M) - the bundle of orthonormal frames on M - , and we have the following commutative diagram:

$$\begin{array}{ccc} O(E) & \xrightarrow{\ \chi_1\ } & O(M) \\ \pi_1 \big\downarrow & & \big\downarrow \chi_2 \\ E & \xrightarrow{\ \pi_2\ } & M \end{array} \quad . \tag{2.1.6}$$

Of course, (2.1.5b) induces the following splitting:

$$TO(E) = \hat{\mathcal{V}} \oplus \hat{\mathcal{H}} \quad , \tag{2.1.7a}$$

$$\hat{\mathcal{V}}_e = \tilde{\delta}'_e(\mathfrak{g}) \quad , \tag{2.1.7b}$$

$$\hat{\mathcal{H}}_e := (\pi'_1)^{-1}(\mathcal{H}_{\pi_1(e)}) \quad , \tag{2.1.7c}$$

with $e \in O(E)$. The connection form on O(E) corresponding to (2.1.7a) will be denoted by ξ .

Now, we observe that formula (2.1.1e) means on $\hat{O}(E)$ that $\hat{\omega}$ is o(1,D-1)-valued. Thus, $\hat{\omega}$ fulfilling (2.1.1a) is completely given by its values on O(E). Since - on the other hand - $\hat{\gamma}$ is constant on O(E), our classification problem is reduced to the problem of characterizing the restriction of $\hat{\omega}$ to O(E). For that purpose we use the decomposition

$$\hat{\omega} = \mathring{\hat{\omega}} + \hat{\varphi} \quad , \tag{2.1.8}$$

where $\mathring{\hat{\omega}}$ is the Levi-Civita connection form corresponding to $\hat{\gamma}$ and $\hat{\varphi}$ is the contorsion form. Then from (2.1.a) and (2.1.3) one gets:

$$\sigma^*_g \mathring{\hat{\omega}} = \mathring{\hat{\omega}} \quad , \tag{2.1.9a}$$

$$\sigma^*_g \hat{\varphi} = \hat{\varphi} \quad . \tag{2.1.9b}$$

Using decomposition (2.1.7a), we see that $\hat{\varphi}$, fulfilling (2.1.9b), is completely characterized by:

i) A mapping

$$\Psi : O(E) \longrightarrow \mathfrak{g}^* \otimes o(1,D-1) \quad , \tag{2.1.10a}$$

$$\Psi(e)(u) := \hat{\varphi}_e(\tilde{\sigma}'_e(u)) \quad , \tag{2.1.10b}$$

$e \in O(E)$, $u \in \mathfrak{g}$, fulfilling

$$\tilde{\sigma}^*_g \Psi = \mathrm{Ad}'g^{-1}(\Psi) \quad , \tag{2.1.10c}$$

with Ad′ denoting the coadjoint representation of G in $\mathfrak{g}^*$.

ii) An o(1,D-1)-valued 1-form on O(M),

$$\tilde{\varphi} : TO(M) \longrightarrow o(1,D-1) \quad , \tag{2.1.11a}$$

given by

$$\tilde{\varphi}_{\chi_1(e)}(X) := \hat{\varphi}_e(X^{\mathcal{H}}) \quad , \tag{2.1.11b}$$

with $X \in T_{\chi_1(e)}O(M)$ and $X^{\mathcal{H}}$ being the horizontal lift of X with respect to ξ .

It remains to analyze $\mathring{\hat{\omega}}$. But, on $\hat{O}(E)$ one has [82]:

$$d\hat{\vartheta} + \mathring{\hat{\omega}} \wedge \hat{\vartheta} = 0 \quad , \tag{2.1.12a}$$

where $\hat{\vartheta}$ denotes the canonical R^D-valued 1-form on $\hat{O}(E)$. The solution of this equation yields:

$$\begin{aligned} &\langle \hat{\vartheta}(Z), \mathring{\hat{\omega}}(Y)\, \hat{\vartheta}(X) \rangle \\ &= -\langle \hat{\vartheta}(Y), d\hat{\vartheta}(Z,X)\rangle + \langle \hat{\vartheta}(Z), d\hat{\vartheta}(X,Y)\rangle + \langle \hat{\vartheta}(X), d\hat{\vartheta}(Y,Z)\rangle \quad , \end{aligned} \tag{2.1.12b}$$

with $X,Y,Z \in TO(E)$, and $\langle \cdot\,,\cdot \rangle$ denoting the standard scalar product on R^D, given by η . Thus, $\mathring{\hat{\omega}}$ is on $O(E)$ completely described by $\hat{\vartheta}$. Decomposing $\hat{\vartheta}$, due to (2.1.7a), we obtain that this quantity is completely characterized by:

i) A mapping

$$\phi : O(E) \longrightarrow \mathfrak{g}^* \otimes R^d \quad , \tag{2.1.13a}$$

$$\phi(e)(u) := \hat{\vartheta}_e(\tilde{\sigma}'_e(u)) \quad , \tag{2.1.13b}$$

$u \in \mathfrak{g}$, satisfying

$$\tilde{\sigma}^*_g \phi = Ad' g^{-1}(\phi) \quad . \tag{2.1.13c}$$

ii) The canonical R^4-valued 1-form ϑ on $O(M)$,

$$\vartheta_{\chi_1(e)}(X) := \hat{\vartheta}_e(X^{\mathcal{H}}) \quad . \tag{2.1.13d}$$

Using the decomposition of the Lie algebra $o(D)$,

$$o(D) = o(1,3) \oplus o(d) \oplus \mathfrak{m} \quad , \tag{2.1.14a}$$

$$\mathfrak{m} := \left\{ \left[\begin{array}{c|c} 0 & u \\ \hline -u^T & 0 \end{array}\right] \quad , \quad u \in L(R^d, R^4) \right\} \quad , \qquad (2.1.14b)$$

and the splitting (2.1.7a), one can decompose $\overset{\circ}{\hat{\omega}}$ on O(E) into six components, and - using formula (2.1.12b) - one can find explicit formulae for these components in terms of geometrical objects just obtained.

For that purpose, it is convenient to use the following vector space isomorphisms:

$$i_e : \hat{\mathcal{V}}_e^* \longrightarrow (R^d)^* \quad , \text{ defined by}$$

$$i_e := (\hat{\vartheta}_{\mathcal{V}}^{-1})^* \quad , \qquad (2.1.15)$$

$$\iota_e : \hat{\mathcal{H}}'^* \longrightarrow (R^4)^* \quad , \text{ defined by}$$

$$\iota_e := (\hat{\vartheta}_{\hat{\mathcal{H}}'}^{-1})^* \quad , \qquad (2.1.16)$$

with $\hat{\mathcal{H}}' = \hat{\mathcal{H}}/\ker \pi'_1$ and $\hat{\vartheta}_{\mathcal{V}}$ and $\hat{\vartheta}_{\hat{\mathcal{H}}'}$ denoting the components of $\hat{\vartheta}_{\restriction O(E)}$ (the restriction of $\hat{\vartheta}$ to O(E)) in the sense of (2.1.7a). Moreover, we have the isomorphism

$$j_e : \overset{2}{\Lambda}\, T_e^* O(E)' \longrightarrow (R^D)^* \wedge (R^D)^* \quad , \quad O(E)' = O(E)/\ker \pi'_1 ,$$

defined by:

$$j_{e \restriction \hat{\mathcal{H}}'\hat{\mathcal{H}}'} := \iota_e \wedge \iota_e \quad , \qquad (2.1.17a)$$

$$j_{e \restriction \hat{\mathcal{H}}'\hat{\mathcal{V}}} := \iota_e \wedge i_e \quad , \qquad (2.1.17b)$$

$$j_{e \restriction \hat{\mathcal{V}}\hat{\mathcal{V}}} := i_e \wedge i_e \quad . \qquad (2.1.17c)$$

Further on we shall write i , ι and j , understanding these mappings pointwise.

Finally, let us choose an adapted orthonormal basis $\{\varepsilon_A\} \equiv \{(\varepsilon_\mu, \varepsilon_\alpha)\}$ in $R^D = R^4 \oplus R^d$, see (2.1.5d), and an orthonormal basis $\{\varepsilon_a\}$ in $\mathfrak{g}$, (with respect to a fixed AdG-invariant scalar product). We denote:

$$\phi(\varepsilon_a) \equiv \phi^\alpha{}_a\, \varepsilon_\alpha \quad , \tag{2.1.18a}$$

$$h_{ab} := \eta_{\alpha\beta}\, \phi^\alpha{}_a\, \phi^\beta{}_b \quad ; \tag{2.1.18b}$$

h_{ab} is a scalar product on $\mathfrak{g}$, induced via ϕ. Moreover, we denote

$$i \circ \mathring{\hat{\omega}} = \omega_{\alpha,}{}^A{}_B\, (\varepsilon^\alpha)^* \otimes (\varepsilon^B)^* \otimes \varepsilon_A \quad , \tag{2.1.19a}$$

$$\iota \circ \mathring{\hat{\omega}} = \omega_{\mu,}{}^A{}_B\, (\varepsilon^\mu)^* \otimes (\varepsilon^B)^* \otimes \varepsilon_A \quad , \tag{2.1.19b}$$

$$j \circ \Xi = 1/2\, \Xi^a{}_{\mu\nu}\, \varepsilon_a \otimes (\varepsilon^\mu)^* \wedge (\varepsilon^\nu)^* \quad , \tag{2.1.19c}$$

with Ξ being the curvature form of ξ ,

$$\iota \circ D\phi = D_\mu \phi^\alpha{}_a\, (\varepsilon^\mu)^* \otimes (\varepsilon^a)^* \otimes \varepsilon_\alpha \quad , \tag{2.1.19d}$$

with $D\phi = d\phi + \mathrm{ad}'\xi(\phi)$ denoting the covariant derivative of ϕ with respect to ξ, and ad' being the coadjoint representation of the Lie algebra,

$$(\widetilde{\mathrm{ad}})^\alpha{}_{\beta\gamma} \equiv \tilde{f}^\alpha{}_{\beta\gamma} = \phi^\alpha{}_a\, f^a{}_{bc}\, (\phi^{-1})^b{}_\beta\, (\phi^{-1})^c{}_\gamma \quad , \tag{2.1.19e}$$

with $\widetilde{\mathrm{ad}}(u,v,w) := \langle u,[v,w]\rangle$ and $f^a{}_{bc}$ being the structure constants of $\mathfrak{g}$ in the basis $\{\varepsilon_a\}$. Finally, we denote the Levi-Civita connection form of the metric γ on M (induced by $\hat{\gamma}$) by $\mathring{\omega}$. In this notation we obtain [99] :

$$\omega_{\mu,\nu\varrho} = (x_1^* \mathring{\omega})_{\mu,\nu\varrho} \quad , \tag{2.1.20a}$$

$$\omega_{\alpha,\nu\varrho} = -1/2\, \phi_{\alpha a}\, \Xi^a{}_{\nu\varrho} \quad , \tag{2.1.20b}$$

$$\omega_{\mu,\alpha\nu} = -\omega_{\mu,\nu\alpha} = -1/2\, \phi_{\alpha a}\, \Xi^a{}_{\mu\nu} \quad , \tag{2.1.20c}$$

$$\begin{aligned} \omega_{\beta,\alpha\mu} = -\omega_{\beta,\mu\alpha} &= 1/2 \left\{ (D_\mu\phi \circ \phi^{-1})_{\alpha\beta} + (D_\mu\phi \circ \phi^{-1})_{\beta\alpha} \right\} \\ &= (D_\mu \phi \circ \phi^{-1})_{(\alpha\beta)} \quad , \end{aligned} \tag{2.1.20d}$$

$$\omega_{\mu,\alpha\beta} = -1/2\left\{(D_\mu\phi\circ\phi^{-1})_{\alpha\beta} - (D_\mu\phi\circ\phi^{-1})_{\beta\alpha}\right\}$$

$$\equiv -(D_\mu\phi\circ\phi^{-1})_{[\alpha\beta]} \quad , \qquad (2.1.20e)$$

$$\omega_{\gamma,\alpha\beta} = -1/2(\tilde{f}_{\gamma\alpha\beta} + \tilde{f}_{\alpha\gamma\beta} + \tilde{f}_{\beta\alpha\gamma}) \quad . \qquad (2.1.20f)$$

Putting

$$\omega = \mathring{\omega} + \tilde{\varphi}^{o(1,3)} \quad , \qquad (2.1.21a)$$

$$\varphi = \tilde{\varphi}^{o(d)\oplus\mathfrak{m}} \quad , \qquad (2.1.21b)$$

where $\tilde{\varphi}^{o(1,3)}$ and $\tilde{\varphi}^{o(d)\oplus\mathfrak{m}}$ are decomposition components - due to (2.1.14a), we get the following classification theorem: Every G-invariant Einstein-Cartan configuration $(\hat{\omega}, \hat{\gamma})$ on E is in 1-1 correspondence with a 5-tupel of geometrical objects

$$(\omega, \xi, \varphi, \phi, \Psi) \quad , \qquad (2.1.22)$$

where ω is the induced Einstein-Cartan connection on O(M), ξ is a G-principal connection on O(E) ⟶ O(M), φ is a 1-form on O(M) with values in $o(d)\oplus\mathfrak{m}$, and ϕ and Ψ are vector-space-valued G-equivariant functions on O(E).

We underline that giving up assumptions i) and ii) leads only to minor modifications. (Removing assumption i) leads - due to the action of the stabilizer H in O(E) - to additional splittings of the matter fields occuring in the above classification theorem, and to the appearance of constraint equations for them. If we remove assumption ii), then it is impossible to perform the last reduction in (2.1.4) in such a way that the bundle O(E) is G-invariant. In this case one has to do the whole analysis presented here - in a completely analogous way - on the bundle $\tilde{O}(E)$.) Some aspects of the general case were discussed in [57] .

2.2. Reduction of the gravitational action

As far as dynamical aspects of Einstein-Cartan theory on a multidimensional universe are concerned, it is not even completely clear, which Lagrangian one should start with. Probably one has to add to the ordinary Einstein-Cartan Lagrangian all non-vanishing terms of Gauss-Bonnet type. Some dynamical considerations - within a special model - were presented, for example, in [100] . In particular, in this model scalar fields giving a potential of Higgs type were obtained. Since in this field only preliminary results exist, we omit a discussion of them in this Review. That means, we restrict ourselves to the torsion free case.Under the assumptions made above, we get immediately: Every G-invariant Riemannian structure on E is in 1-1 correspondence with a triple

$$(\mathring{\omega}, \xi, \phi) \quad . \tag{2.2.1}$$

This coincides with the classification result of Coquereaux and Jadczyk, cited in section 1.2., if we put there $H=\{\mathbb{1}_G\}$. (The field ϕ can be replaced by a field h_{ab} with values in scalar products on $\mathfrak{g}$, see (2.1.18b).)

The natural generalization of the Einstein-Cartan Lagrangian [94] to the D-dimensional universe is

$$L = \frac{1}{(D-2)!} \varepsilon_{A_1 \ldots A_D} \hat{\vartheta}^{A_1} \wedge \ldots \wedge \hat{\vartheta}^{A_{D-2}} \wedge \Omega^{A_{D-1}A_D} \quad , \tag{2.2.2a}$$

$$\eta^{-1} \cdot \Omega = 1/2\, \Omega^{AB} \varepsilon_A \wedge \varepsilon_B \quad , \tag{2.2.2b}$$

with Ω denoting the curvature form of the metric linear connection. Obviously, we have

$$L = \Omega^{AB}{}_{AB}\, \hat{\vartheta}^1 \wedge \ldots \wedge \hat{\vartheta}^D \quad , \tag{2.2.2c}$$

with

$$j \circ \eta^{-1} \circ \Omega = 1/4\, \Omega^{AB}{}_{CD}\, \varepsilon_A \wedge \varepsilon_B \otimes (\varepsilon^C)^* \wedge (\varepsilon^D)^* \quad . \tag{2.2.2d}$$

In the torsion free case, Ω is the curvature form of $\mathring{\omega}$, and it is easy to show that

$$L = \pi_1^* (\hat{R} \cdot v_E) \quad , \tag{2.2.3}$$

where $\hat{R}$ is the scalar curvature of $\hat{\gamma}$ and v_E is the canonical volume form on E. Thus, for the torsion free case, L is the lift of the ordinary Einstein-Hilbert Lagrangian to the bundle space. For simplicity of notation we omit the overall factor $1/16\pi G$ in L. Using

$$\Omega = d\hat{\mathring{\omega}} + 1/2[\hat{\mathring{\omega}}, \hat{\mathring{\omega}}] \quad , \tag{2.2.4}$$

and formulae (2.1.20a) - (2.1.20f), denoting the scalar curvature of $\mathring{\omega}$ by R_M, the scalar curvature of $\phi^*\eta$ by $R_G(\phi)$ (function on O(E) with values in scalar curvatures) and

$$\nabla := d + ad\xi + ad\,\hat{\mathring{\omega}}^{o(1,3)} \quad , \tag{2.2.5}$$

we get:

$$\Omega^{\mu\nu}{}_{\mu\nu} = \pi_1^*(R_M) - 3/4\, h_{ab} \Xi^a_{\mu\nu} \Xi^{b\,\mu\nu} \quad , \tag{2.2.6a}$$

$$\Omega^{\alpha\mu}{}_{\alpha\mu} = -\nabla_\mu((D^\mu\phi\cdot\phi^{-1})^\alpha{}_\alpha) + 1/4\, h_{ab} \Xi^a_{\mu\nu} \Xi^{b\,\mu\nu}$$
$$- (D_\mu\phi\cdot\phi^{-1})^{(\alpha\beta)}(D^\mu\phi\cdot\phi^{-1})_{(\alpha\beta)} \quad , \tag{2.2.6b}$$

$$\Omega^{\alpha\beta}{}_{\alpha\beta} = R_G(\phi) - (D_\mu\phi\circ\phi^{-1})^\alpha{}_\alpha(D^\mu\phi\circ\phi^{-1})^\beta{}_\beta$$
$$+ (D_\mu\phi\circ\phi^{-1})^{(\alpha\beta)}(D^\mu\phi\cdot\phi^{-1})_{(\alpha\beta)} \quad . \tag{2.2.6c}$$

This yields:

$$\Omega^{AB}{}_{AB} = \pi_1^*(R_M) + R_G(\phi) - 1/4\, \eta_{\alpha\beta}\phi^\alpha_a \phi^\beta_b \Xi^a_{\mu\nu} \Xi^{b\,\mu\nu}$$
$$- (D_\mu\phi\cdot\phi^{-1})^\alpha{}_\alpha(D^\mu\phi\cdot\phi^{-1})^\beta{}_\beta - (D_\mu\phi\circ\phi^{-1})^{(\alpha\beta)}(D^\mu\phi\circ\phi^{-1})_{(\alpha\beta)}$$
$$-2\nabla_\mu((D^\mu\phi\cdot\phi^{-1})^\alpha{}_\alpha) \quad . \tag{2.2.7a}$$

Observe that ∇ acting on $(D^\mu\phi\circ\phi^{-1})^\alpha{}_\alpha$ coincides with $\tilde{\nabla} = d + ad\,\hat{\mathring{\omega}}^{o(1,3)}$.

A simple calculation - using (2.1.18b) - gives:

$$\Omega^{AB}{}_{AB} = \pi_1^*(R_M) + R_G(h) - 1/4\, h_{ab}\, \Xi^a_{\mu\nu}\, \Xi^{b\mu\nu}$$

$$- 1/4(h^{ab}D_\mu h_{ab})(h^{cd}D^\mu h_{cd}) - 1/4\, h^{ab}h^{cd}D_\mu h_{ac}D^\mu h_{bd}$$

$$- \nabla_\mu(h^{ab}D^\mu h_{ab}) \quad . \qquad (2.2.7b)$$

If we pass to a description in terms of local D-bein fields, we obtain from (2.2.7b) the result of Cho and Freund [70]. The result coincides also with that of Coquereaux and Jadczyk [71], if we put there $H=[0_G]$.

For the lift of the volume form appearing in (2.2.2) we get

$$\hat{\vartheta}^1 \wedge \ldots \wedge \hat{\vartheta}^D$$

$$= \chi_1^*(\vartheta^1 \wedge \ldots \wedge \vartheta^4)(\det(h))^{1/2}(\tilde{\zeta}^*)^{-1}(\vartheta_G) \quad , \qquad (2.2.8a)$$

with ϑ denoting the canonical R^4-valued 1-form on O(M) and ϑ_G being the value of the canonical volume form on G (with respect to the fixed AdG-invariant scalar product), at the identity. In the orthonormal basis $\{\varepsilon_a\}$ we have

$$\vartheta_G = (\varepsilon^1)^* \wedge \ldots \wedge (\varepsilon^d)^* \quad , \qquad (2.2.8b)$$

$$(\tilde{\zeta}^*)^{-1}\vartheta_G = (\tilde{\varepsilon}^1)^* \wedge \ldots \wedge (\tilde{\varepsilon}^d)^* \quad , \qquad (2.2.8c)$$

where $\{\tilde{\varepsilon}_a\}$ denote the Killing fields generated by $\{\varepsilon_a\}$, and $(\tilde{\varepsilon}^a)^*$ are the corresponding dual 1-forms.

The Lagrangian (2.2.7) describes a unified theory of the gravitational field, a Yang-Mills field and a set of scalar fields on reduced space time M. In the kinetic terms (4th and 5th term in (2.2.7)), the scalar fields are minimally coupled to the Yang-Mills field. However, the pure Yang-Mills part (3rd term) contains an additional, non-minimal coupling with the scalar fields. The second term is a self-interaction scalar potential. A systematic analysis of its structure (including the case of non-trivial stabilizer of the G-ction on E) seems to be quite complicated and has not yet been done. For a discussion of some aspects of this question see [71,101], and further references therein. In [71] one can find examples showing that - usually - $R_G(\phi)$ is not bounded from below, indicating that the ordinary Higgs mechanism does not work within this framework.

If we assume that ϕ is constant on O(E), then (2.2.7) describes an Einstein-Yang-Mills system with cosmological constant.

2.3. Spontaneous compactification and dimensional reduction

One of the questions concerning the dimensional reduction theory is, whether the multidimensional universe of the form

$$E = M \times G/H \qquad (2.3.1)$$

should be given a priori, or rather should be obtained by a dynamical mechanism - called spontaneous compactification. Unfortunately, as already mentioned in the introduction, starting from Einsteins theory on E one cannot find solutions of the field equations giving (2.2.1), with G being non-Abelian. For that purpose one has to add matter to the multidimensional theory. Usually, one adds Yang-Mills fields [8,12-18,21,22,72-75,102,103]. This seems to be a rather serious modification of the original Kaluza-Klein idea. Therefore, one may wonder, whether it is possible to find interesting solutions (2.2.1) by generalizing the space time structure, e.g. by adding torsion [97] or by considering even more general structures [33]. Another possibility is to add higher order curvature terms to the Einstein-Hilbert Lagrangian, see e.g. [104] .

We underline that the concept of spontaneous compactification also plays an important role in theories of supergravity [23,105-107] and superstrings [28] .

In this Review we concentrate on spontaneous compactification in the framework of Einstein-Yang-Mills theories, and especially on the relation between spontaneous compactification and dimensional reduction [72-75] . It turns out that the multidimensional solutions can be - on one hand - interpreted in terms of fields appearing in the reduced theory; on the other hand, simple solutions of the reduced theory (extrema of the Higgs potential) can be used to build up multidimensional field configurations of spontaneous compactification type.

In most of the papers known to us the internal space G/H was supposed to be symmetric. In recent years, however, extensions to the non-symmetric case were also sudied [21,22,75,102,103] .

2.3.1. The equations of spontaneous compactification

We start with a theory on a (pseudo)-Riemannian manifold E, dimE = D = 4+d, given by the following action :

$$S = S_{EH} + S_{YM} \quad , \tag{2.3.2a}$$

$$S_{EH} = \int_E d\vartheta_E \left(\frac{1}{16\pi\varkappa} \hat{R} - \Lambda_{cosm} \right) \quad , \tag{2.3.2b}$$

$$S_{YM} = -\int_E d\vartheta_E \, \frac{1}{8g^2} \langle \hat{F}_{MN}, \hat{F}^{MN} \rangle_{k} \quad , \tag{2.3.2c}$$

where $\hat{R}$ denotes the scalar curvature on E, Λ_{cosm} the comological constant of the multidimensional theory and $d\vartheta_E = d^Dx\sqrt{-\det\hat{\gamma}}$, for a coordinate system $\{x^M\}$ on E. (For the purpose of this section it is more convenient to work from the very beginning with the Lagrangian on E, and not on the bundle space.)

We look for solutions to the field equations corresponding to (2.3.2), within the following class of configurations:

i) The metric on E is of the form

$$\hat{\gamma} = \eta \oplus \gamma \quad , \tag{2.3.3a}$$

where η is the Minkowski metric on M and γ denotes a G-invariant metric on G/H.

ii) The gauge potential on M vanishes.
iii) The gauge potential on G/H is G-symmetric.

For this ansatz the Einstein-Yang-Mills equations on E take the following form [12-14] :

$$\hat{R}_{ab} = 4\pi \tilde{\varkappa} \tilde{g}^{-2} \langle \hat{F}_{ac}, \hat{F}_b{}^c \rangle_{k} \quad , \qquad (2.3.4a)$$

$$\nabla_a F^{ab} + [\hat{A}^a, \hat{F}^{ab}] = 0 \quad , \qquad (2.3.4b)$$

$$\Lambda_{cosm} = 1/8\, g^{-2} \langle \hat{F}_{ab}, F^{ab} \rangle_{k} \quad , \qquad (2.3.4c)$$

where all geometrical objects appearing in (2.3.4) are written down in a chosen (local) coordinate system $\{y^a\}$ on G/H. In particular,

$$\nabla_a \hat{F}^{ab} = \frac{\partial}{\partial y^a} \hat{F}^{ab} + \Gamma^l_{al} \hat{F}^{ab} \quad , \qquad (2.3.4d)$$

with Γ^l_{ik} denoting the Christoffel symbols corresponding to γ_{ab}.

First, we observe that - due to assumptions i) - iii) - the system (2.3.4) reduces to a system of nonlinear algebraic equations. Indeed, since the gauge potential $\hat{A}$ and the metric γ are G-symmetric, it is sufficient to solve (2.3.4) at one point; the whole solution is then found by G-invariant extension to G/H. More precisely: From G-invariance of γ and $\hat{A}$ follows that γ_{ab}, $R_{ab}(\gamma)$ and $T_{ab}(\gamma, \hat{A}) = \langle \hat{F}_{ac}, \hat{F}_b{}^c \rangle_{k}$ are G-invariant symmetric tensor fields of second rank on G/H. It is well known [82] that the space of these tensors is in 1-1 correspondence with the space of ad$\mathfrak{h}$ -invariant symmetric bilinear forms on $\mathfrak{m}$. Here we consider again the case, when assumptions i) - iv) , made in subsection 1.3.1., are fulfilled. Then the metric is of the form (1.3.8), and chosing an orthonormal basis $\{v_a\} \equiv \{v_{(k,r)}\}$ in $\mathfrak{m}$, fulfilling (1.3.9b), we have for $\gamma_{ab} := \gamma(v_a, v_b)$:

$$\gamma_{(k,r),(k',s)} = m_k^{-2} \delta_{kk'} \delta_{rs} \quad . \qquad (2.3.5a)$$

In an analogous way we get:

$$\hat{R}_{(k,r),(k',s)} = \hat{R}_k(m_1^2)\,\delta_{kk'}\,\delta_{rs} \quad , \qquad (2.3.5b)$$

$$T_{(k,r),(k',s)} = T_k(m_1^2,A)\,\delta_{kk'}\,\delta_{rs} \quad . \qquad (2.3.5c)$$

Now we adopt the general form of G-invariant gauge potentials, see (1.2.17b), to our case. We get

$$\hat{A} = \tau'(\bar{\Theta}^{\gamma}) + \phi(\bar{\Theta}^{m}) \quad , \qquad (2.3.6a)$$

with ϕ not depending on $x \in M$, and - according to (1.2.18b):

$$F = 1/2\,\{[\phi(\bar{\Theta}^{m}),\phi(\bar{\Theta}^{m})] - \phi([\bar{\Theta}^{m},\bar{\Theta}^{m}]_{|m}) - \tau'([\bar{\Theta}^{m},\bar{\Theta}^{m}]_{|\gamma})\}. \qquad (2.3.6b)$$

Since ϕ does not depend on $x \in M$, the functions $\phi_s^{(k,i)}$ in (1.3.7) are constants in the case under consideration, and the spontaneous compactification equations are in effect - as already mentioned - non-linear algebraic equations for the parameters m_k^2 and $\phi_s^{(k,i)}$.

2.3.2. Solving the equations of spontaneous compactification

Due to (2.3.5), equation (2.3.4a) takes the following form:

$$\hat{R}_k(m_1^2) = 4\pi\varkappa\cdot g^{-2}\,T_k(m_1^2,\hat{A}) \quad , \quad k = 0,\dots,N \ , \qquad (2.3.7)$$

where $\varkappa\cdot g^{-2} = \tilde{\varkappa}\cdot\tilde{g}^{-2}$. The quantities $\hat{R}_k$ (treated as functions of the parameters m_1) can be calculated using standard formulae [82] . In order to calculate the right-hand-side of (2.3.7) we have to find $\hat{A}$. For that purpose we have to solve equation (2.3.4b). Inserting (2.3.6a) into (2.3.4b), calculating the Christoffel symbols for γ given by (2.3.5a) and using the Maurer-Cartan equations for $\bar{\Theta}$, we get

$$1/2\ \gamma_{cb}\gamma^{ef}\gamma^{da} f^{c}{}_{df}\,\hat{F}_{ae} + \gamma^{ad}[\phi_d,\hat{F}_{ab}] = 0 \quad . \qquad (2.3.8)$$

Here we again denoted $\phi_a \equiv \phi(v_a)$; $\hat{F}_{ab}$ is given by

(1.2.19c) and $f^{a}{}_{bc}$ are the structure constants of $\mathfrak{g}$ in the basis $\{v_a\}$. Equation (2.3.8) was - for isotropy irreducible spaces G/H - obtained in [21].

We see that $\phi = 0$ is, in general, not a solution of (2.3.8). However, if among the representations of $\mathrm{ad}\,\mathfrak{h}(\mathfrak{m})$ are no representations equivalent to $\mathrm{ad}\,\mathfrak{h}$, then - as one easily shows - equation (2.3.8) has a trivial solution.

To solve (2.3.8), we use the dimensional reduction method. One can show that every extremum of the action (2.3.2) in the class of G-symmetric fields is an extremum in the class of all fields [36,108]. On the other hand, it is easy to convince oneself that equation (2.3.4b) is obtained by variation of

$$\tilde{S} = 1/8\ \tilde{g}^{-2} \int_{G/H} d\vartheta_{G/H} \langle \hat{F}_{ab}, \hat{F}^{ab} \rangle_{k} \quad . \qquad (2.3.9)$$

But, restricted to symmetric fields, the function appearing in this integral is constant on G/H. Thus, one can calculate it at a given point, say $[\mathbb{1}_G] \in G/H$, and integrate over G/H. The result is, obviously, the scalar field potential $V(\phi)$ of the reduced theory, see (1.2.19). We conclude that equation (2.3.4b) is equivalent to the equation for the extrema of the scalar field potential (with the variation to be performed in accordance with constraint equation (1.3.3)).

To summarize, we have the following algorithm for finding the solutions to equations (2.3.4):

i) Using the dimensional reduction method, we calculate the scalar field potential of the reduced Yang-Mills theory, described by the action (2.3.2c), for G-symmetric fields. The potential is a function of ϕ and of the parameters m_k^2.

ii) For fixed parameters m_k, we find the extrema $\tilde{\Phi} = \tilde{\Phi}(m_k^2)$ of $V(\phi)$.

iii) Next we calculate $T_k(m_k^2, \tilde{\Phi})$ and insert these quantities into (2.3.7). The result is a system of non-linear, algebraic equations for m_k^2. It is clear from (2.3.7) that the number of equations coincides with the number of parameters.

(Finally, one has to calculate the right-hand-side of (2.3.4c), to obtain the value of the cosmological constant. We shall comment on problems related to this subject in subsection 2.3.3.)

First, let us demonstrate this method for the case of G/H being symmetric. Let us assume - for simplicity - that the scalar field potential is of the form

$$V(\phi) = m^4 \left\{ c_1 (\|\phi\|^2 - c_2/c_1)^2 + c_o \right\} \quad , \tag{2.3.10}$$

see (1.3.28b), with c_i being dimensionless constants and m^{-1} characterizing the "size" of G/H. The extrema of (2.3.10) are given by:

$$\tilde{\phi}_1 = 0 \quad , \tag{2.3.11a}$$

$$\|\tilde{\phi}_2\|^2 = c_2/c_1 \quad . \tag{2.3.11b}$$

(In this simple case, $\tilde{\phi}_i$'s do not depend on m .)
From (2.3.5c) we have:

$T(m^2, \hat{A}) = d^{-1} \cdot m^{-2} \, V(\phi)$. Therefore:

$$T_1 := T(m^2, \hat{A})\restriction_{\phi = \tilde{\phi}_1} = d^{-1} \cdot m^{-2} \, V(0) = d^{-1} m^2 (c_2^2/c_1 + c_o) \, , \tag{2.3.12a}$$

$$T_2 := T(m^2, \hat{A})\restriction_{\phi = \tilde{\phi}_2} = d^{-1} \cdot m^{-2} \, V(\tilde{\phi}_2) = d^{-1} m^2 c_o , \tag{2.3.12b}$$

The left-hand-side of (2.3.7) is in this case given by

$$\hat{R}(m^2) = d^{-1} m^2 \cdot R_{G/H} \quad , \tag{2.3.12c}$$

with $R_{G/H}$ denoting the scalar curvature of G/H. One has $R_{G/H} = r \cdot m^2$, with r being a dimensionless parameter characterizing G/H. Thus,

$$\hat{R}(m^2) = d^{-1} r \quad . \tag{2.3.12d}$$

Finally, we get two solutions to equations (2.3.7) and (2.3.8):

$$m_1^2 = \frac{g^2 r}{4\pi\varkappa} \cdot \frac{1}{c_2^2/c_1 + c_o} \quad , \quad \tilde{\phi}_1 = 0 \; , \; \hat{A}_1 = \tau^{\backslash}(\bar{\Theta}^{\gamma}) \; , \tag{2.3.13a}$$

$$m_2^2 = \frac{g^2 r}{4\pi\varkappa} \cdot \frac{1}{c_0} \quad , \|\tilde{\phi}_2\|^2 = c_2/c_1 \ , \ \hat{A}_2 = \tau^{\backprime}(\overline{\Theta}\mathfrak{h}) + \tilde{\phi}_2(\overline{\Theta}^{\mathfrak{m}}) \ .$$ (2.3.13b)

In most of the papers known to us [12-20] , solutions of type (2.3.13a) were discussed. Those solutions correspond to the so called canonical invariant connection [82] $\Theta^{\mathfrak{h}}$ on G ⟶ G/H, as visible from (2.3.13a). Obviously, this class of solutions is given by the (unstable) local maximum $\phi = 0$ of the Higgs potential (one easily shows that $c_2 > 0$.)

The configuration (2.3.13b) corresponds to the stable Higgs vacuum of the reduced theory and, therefore, should be probably considered to be more adequate from the physical point of view.

If we had not used the dimensional reduction method, we would have had to solve (2.3.8) directly. In the case under consideration it reduces to

$$\sum_a [\phi_a,[\phi_b,\phi_a] + \tau^{\backprime}([v_a,v_b])] = 0 \quad , \qquad (2.3.14)$$

because $\gamma_{ab} = m^{-2}\delta_{ab}$. A straightforward calculation shows that this equation has the two solutions (2.3.11). The authors of [15-17] propose to solve - instead of (2.3.4b) - a (stronger) equation:

$$\nabla_a \hat{F}_{bc} + [\hat{A}_a,\hat{F}_{bc}] = 0 \quad . \qquad (2.3.15a)$$

This equation is, obviously, a parallizibility condition for $\hat{F}$. For symmetric spaces (2.3.15a) takes the form:

$$[\phi_a,[\phi_b,\phi_c] - \tau^{\backprime}([v_b,v_c])] = 0 \ . \qquad (2.3.15b)$$

Assuming $\mathfrak{N}^{(i)} = \{0\}$ in (1.3.6b), one can show that (2.3.14) and (2.3.15b) are equivalent in the following cases:

i) $\mathfrak{h}$ is simple,

ii) $\mathfrak{h} = \mathfrak{h}_1 \oplus \mathfrak{h}_2$ is semisimple and the indices λ_1 and λ_2 are equal to each other,

iii) $[\phi(\mathfrak{V}),\phi(\mathfrak{V})] = 0$.

In all other cases (2.3.15b) has only the trivial solution.

Finally, let us consider an example with G/H being non-symmetric:

$$G/H = Sp(2)/(SU(2) \times U(1)) \quad , \qquad (2.3.16a)$$

$$K = SU(5) \quad . \qquad (2.3.16b)$$

Decomposing $ad_{\mathfrak{h}}(\mathfrak{m})$ into irreducible components, we get:

$$ad_{\mathfrak{h}}(\mathfrak{m}) = [\underline{1}](2) \oplus [\underline{1}](-2) \oplus [\underline{2}](1) \oplus [\underline{2}](-1) \quad , \qquad (2.3.16c)$$

Thus, γ is characterized by two parameters, m_1 and m_2. Denoting $\mu = m_2^2/m_1^2$, we have:

$$\hat{R}_1 = 3 - 1/2\mu \ ,$$

$$\hat{R}_2 = 2 + 1/2\mu \ ,$$

for the components of the scalar curvature in (2.3.5b).

Let us assume that

$$\tau : SU(2) \times U(1) \longrightarrow SU(5)$$

is such that the scalar field of the reduced theory is in the fundamental representation of the non-Abelian component of the reduced gauge group C = SU(3) x U(1) [64]. Applying our method described above, we get:

$$V(\phi) = m_1^4 \left\{ \tfrac{12}{5}(3+\mu^2) - 4(2-\mu)\|\phi\|^2 + 3\|\phi\|^4 \right\} \quad , \qquad (2.3.17a)$$

$$T_1 = m_1^2 \left\{ \tfrac{3}{4}\|\phi\|^4 - 2\|\phi\|^2 + \tfrac{9}{5} + \tfrac{1}{2}\mu\|\phi\|^2 \right\} \quad , \qquad (2.3.17b)$$

$$T_2 = m_1^2 \left(\|\phi\|^2 + \tfrac{6}{5}\mu \right) \quad . \qquad (2.3.17c)$$

(The result of our calculation differs from the result obtained in [64] in the o-th order term of $V(\phi)$!)

Let us consider two cases:

a) $\mu \geq 2$:
Then $V(\phi)$ has an absolute minimum for $\phi = 0$.

b) $\mu < 2$:

Then $V(\phi)$ has two extrema; namely one local maximum for $\tilde{\Phi}_1 = 0$ and an absolute minimum for $\|\tilde{\Phi}_2\|^2 = 2/3(2-\mu)$.

Finally, we get two solutions to the equations of spontaneous compactification:

$$m_1^2 \simeq \frac{g^2}{4\pi\varkappa}\cdot 1{,}4 \ , \quad \mu \simeq 1{,}312 \ , \quad \tilde{\phi} = 0 \ , \quad \hat{A} = \tilde{\tau}(\bar{\Theta}^{\mathfrak{h}}) \ , \qquad (2.3.18a)$$

(unstable local maximum of the Higgs potential) ,

$$m_1^2 = \frac{g^2}{4\pi\varkappa}\cdot\frac{5}{2} \ , \quad \mu = 1/2 \ , \|\tilde{\phi}\|^2 = 1, \ \hat{A} = \tilde{\tau}(\bar{\Theta}^{\mathfrak{h}}) + \tilde{\phi}(\bar{\Theta}^{\mathfrak{m}}) \ . \qquad (2.3.18b)$$

We see that in both cases $\mu < 2$, i.e. field dynamics chooses the solution belonging to the sector where we have spontaneous symmetry breaking.

<u>2.3.3. Problems related to spontaneous compactification</u>

a) The first problem one has to discuss is that of the classical stability of compactifying solutions. For that purpose, one has to investigate the spectrum of vacuum fluctuations; the modes with negative mass square (tachions) lead to instability. If we have a model leading to a non-trivial matter field multiplet ϕ , then - of course - solutions corresponding to the maximum of the potential $V(\phi)$ are unstable, at least with respect to fluctuations from the symmetric sector [19,37] .

If the matter field multiplet is trivial, $\phi \equiv 0$ (there are no equivalent representations in $\mathfrak{m}$ and $\mathfrak{k}$), then the solution $\phi = 0$ is stable [19,20] ; but the reduced theory is not attractive from the physical point of view.

In the general case, if we have a non-trivial matter field multiplet, then the configurations corresponding to the minimum of $V(\phi)$ seem to be natural candidates for stable vacua. This conjecture was first formulated in [72], see also [73,74] . Concrete examples studied in [72] confirm this conjecture. Nevertheless, in the general case, this question is still open.

Finally, we make a few remarks on the case, when the reduction gives a non-trivial multiplet and the solution $\phi \neq 0$, corresponding to a minimum of $V(\phi)$, extends $\tau' : \mathfrak{h} \longrightarrow \mathfrak{k}$ to a homomorphism $\Lambda : \mathfrak{g} \longrightarrow \mathfrak{k}$. Then $V(\phi)=0$ and the equations of spontaneous compactification have no meaningful solutions; one gets vanishing scalar curvature on G/H.

b) It turns out that gravitational fluctuations usually do not give considerable contributions to vacuum fluctuactions [19,72]; see, however, the first paper in [20] !

Classical stability as discussed above does not guarantee stability on the quantum level [77] . This problem is closely related to the approach, where spontaneous compactification is a result of quantum effects [109] . Here we shall not discuss these questions.

c) If one tries to construct realistic models using the dimensional reduction method, (e.g. the bosonic sector of the Georgi-Glashow model), then one finally has to fit the parameters of the theory (the multidimensional coupling constant g and the parameters m_k) to observable quantities. This gives values for m_k of the order of masses of particles appearing in the model under consideration. On the other hand, the solutions of equations of spontaneous compactification give values for m_k of the order of the Planck's length. This is a contradiction, which - at least on the classical level - cannot be overcome.

d) Due to (2.3.4c), we get for the cosmological constant

$$\Lambda_{cosm} = -1/8\, \tilde{g}^{-2}\, V(m, \tilde{\Phi}) \quad ,$$

with m and $\tilde{\Phi}$ being solutions of (2.3.4a) and (2.3.4b). This is a fine tuning, which - obviously - is model dependent and, therefore, seems to be quite unnatural. On the other hand, if one puts $\Lambda_{cosm} = 0$ on the level of the multidimensional theory, then one gets a cosmological constant of the 4-dimensional theory, which is many orders of magnitude greater than the observed upper limit, (cosmological constant puzzle).

3. Dimensional reduction of matter fields and model building

3.1. Dimensional reduction of matter fields

If one is interested in constructing by dimensional reduction realistic models on physical space time containing, in particular, fermionic fields, one has to add the latter to the multidimensional theory. Thus, one has to solve the problem of their dimensional reduction. The appropriate geometrical language to describe fermions is that of spin structures [110] . However, whereas in the case of frame bundles group actions on the base manifold lift naturally to the bundle space, for spin bundles this question is much more delicate [111]. In particular, a classification of spin bundles admitting lifts of, say, simple group actions is not known to us. Therefore, we decided - as usually done, at least in the physical literature - to treat spinor fields as fields in a double-valued representation of the (pseudo)-orthogonal group. If one assumes the existence of a lift, then one can perform reductions of spin bundles [57], similar to frame bundle reductions discussed in this sectio: compare also with section 2.1.

We denote again - as in section 2.1. - by L(E) the bundle of linear frames on E, and by $\hat{O}(E)$ the subbundle of orthonormal frames. The sum [82], $\hat{Q}$, of these two bundles is given by

$$\hat{Q} := \left\{ (e,\hat{p}) \in \hat{O}(E) \times \hat{P} : \ \pi_O(e) = \pi_{\hat{P}}(\hat{p}) \right\} \quad , \qquad (3.1.1)$$

with π_O and $\pi_{\hat{P}}$ denoting the canonical projections in $\hat{O}(E)$ and $\hat{P}$. Of course, $\hat{Q}$ is a principal bundle over E with structure group $\hat{S} = O(1,D-1) \times K$. We denote the canonical projection in $\hat{Q}$ by π and the right action of $\hat{S}$ by ψ .

A matter field of arbitrary spin-tensor type is an equivariant mapping

$$\hat{\varsigma} \ : \ \hat{Q} \longrightarrow \mathcal{R}_{\mathcal{D}} \ , \qquad (3.1.2a)$$

$$\hat{\varsigma} \circ \psi_s = \mathcal{D}\,(s^{-1}) \circ \hat{\varsigma} \quad , \qquad (3.1.2b)$$

with $s \in \hat{S}$ and $\mathcal{R}_{\mathcal{D}} = \mathcal{R}_{\Delta} \otimes \mathcal{R}_{\alpha}$, $\mathcal{D} = (\Delta , \alpha)$, being the representation space of $\hat{S}$ (here Δ is a representation of O(1,D-1) and α denotes a representation of the gauge group K).

Now, let us assume that a simple group action δ of G on E is given and that $\hat{P}$ is a bundle admitting a lift - as discussed in section 1.1. Moreover, we assume that the metric $\hat{\gamma}$ on E is G-invariant. Since δ lifts naturally to the bundle $\hat{O}(E)$, we have an action of G on $\hat{Q}$:

$$\delta : G \times \hat{Q} \longrightarrow \hat{Q} \quad , \qquad (3.1.3a)$$

$$\delta_g \in \mathrm{Aut}(\hat{Q}) \text{ , for all } g \in G \text{ .} \qquad (3.1.3b)$$

A matter field $\hat{\varphi}$ of type $\mathcal{D}$ is called G-invariant if and only if

$$\delta_g^* \hat{\varphi} = \hat{\varphi} \quad . \qquad (3.1.4)$$

For a given (local) section $s: E \longrightarrow \hat{Q}$, the corresponding local representative $s^* \hat{\varphi}$ will be G-symmetric.

Let there be chosen - as in section 1.1.- a section $s: M \longrightarrow E$, $s(M) \equiv \tilde{M}$, fulfilling (1.1.3b). Then we build $\tilde{Q} := \pi^{-1}(\tilde{M})$,

$$\tilde{Q} = \tilde{O}(\tilde{M},O(1,D-1) + \tilde{P}(\tilde{M},K) \quad , \qquad (3.1.5)$$

and observe that a G-invariant matter field $\hat{\varphi}$ is completely determined by its values on $\hat{Q}$. Moreover, see (1.1.11), it fulfills

$$\tilde{\delta}_h^* \tilde{\varphi} = \tilde{\varphi} \quad , \qquad (3.1.6)$$

with $\tilde{\varphi} \equiv \hat{\varphi} \restriction \tilde{Q}$ and $h \in H$.

We take $\hat{\gamma}$ of the form (1.2.3b). Observe that $\hat{\gamma}$ and the action of G on E induce a sequence of natural fibre bundle reductions, analogous to (2.1.4):

$$\tilde{O}(\tilde{M}) \longrightarrow \tilde{O}'(\tilde{M}) \longrightarrow O(\tilde{M}) \quad , \qquad (3.1.7)$$

with $\tilde{O}'(\tilde{M}) \equiv \tilde{O}'(\tilde{M},O(1,3) \times O(d))$, $4+d = D$, denoting the bundle of adapted orthonormal frames on E , restricted to $\tilde{M}$, and $O(\tilde{M})$ being the bundle of orthonormal frames on $\tilde{M}$.(The second reduction is obtained by the choice of an orthonormal basis in $T_{[e]}G/H$.)

The first reduction in (3.1.7) induces a reduction of $\tilde{Q}$ to a bundle $\tilde{Q}'$ with structure group O(1,3)x O(d)xK. Since $\hat{\gamma}$ is G-invariant, $\tilde{Q}'$ is H-invariant and there exists - analogously to (1.1.6) - a mapping

$$\beta : H \times \tilde{Q}' \longrightarrow O(d) \times K \quad , \qquad (3.1.8)$$

which for every $\tilde{q} \in \tilde{Q}'$ is a group homomorphism. Of course, β is built up from τ ,defined by (1.1.6), and a mapping

$$\lambda : H \times \tilde{O}'(\tilde{M}) \longrightarrow O(d) \quad ,$$

defined analogously to τ . We underline that - whereas τ can be chosen freely - λ is induced by the isotropy representation of H on the space tangent to the orbit.

Similarly to (1.1.7a) we have - for a given point $\tilde{q}_o \in \tilde{Q}'$ - the subbundle

$$Q' := \left\{ \tilde{q} \in \tilde{Q}' : \quad \beta\,(\tilde{q},h) = \beta\,(h)\ ,\ h \in H \right\} \quad , \qquad (3.1.9)$$

$\beta(h) \equiv \beta(\tilde{q}_o,h)$, with structure group $S' = O(1,3) \times C_{O(d)}(\lambda(H)) \times C$, where C denotes again the centralizer of $\tau(H)$ in K.

Finally, the second reduction in (3.1.7) enables us to reduce Q' to a subbundle Q with structure group S = O(1,3)xC. We have $Q = O(\tilde{M}) + P(\tilde{M},C)$. Because of (3.1.2b), $\tilde{\varsigma}$ is completely characterized by its values on Q, which we denote by ς , $\varsigma \equiv \tilde{\varsigma}\restriction_Q$. On this subbundle (3.1.6) takes the form

$$\mathcal{D}\,(\beta\,(h^{-1}))\,\varsigma \;=\; \varsigma \quad , \text{ for all } h \in H \ . \qquad (3.1.10)$$

This means that ς takes values in the subspace $\mathcal{R}_o$ of the trivial representation of $\beta(H)$ in $\mathcal{R}_{\mathcal{D}}$. Since $\beta(H)$ by definition of S' commutes with $S \subset S'$, this space is invariant under the action of the reduced structure group S. Thus, we have found that a G-invariant matter field $\hat{\varsigma} : \hat{Q} \longrightarrow \mathcal{R}_{\mathcal{D}}$ is in 1-1 correspondence with a matter field $\varsigma : Q \longrightarrow \mathcal{R}_o$, on a bundle over reduced space time $\tilde{M}$ with structure group O(1,3)xC, taking values in the space $\mathcal{R}_o$ of the trivial representation of H in $\mathcal{R}_{\mathcal{D}}$. (All reconstructions discussed in section 1.1. can be performed here analogously.)

Now we shall discuss the reduction of the physical action for fermions, which means we take Δ to be a spinor representation of O(1,D-1). Following [25], (but see also [116]), we consider the Clifford algebra of Dirac matrices associated with E,

$$[\hat{\Gamma}^A, \hat{\Gamma}^B]_+ = 2\eta^{AB} \quad , \tag{3.1.11a}$$

$$\hat{\Gamma}^A = \tilde{\Gamma}^A \otimes 1\!\!1 \quad , \qquad 0 \leq A \leq 3 \quad , \tag{3.1.11b}$$

$$\hat{\Gamma}^A = \tilde{\Gamma}^5 \otimes \Gamma^{A-3} \quad , \qquad 4 \leq A \leq D-1 \quad , \tag{3.1.11c}$$

with $\eta^{AB} = \mathrm{diag}(-1,1,\ldots,1)$ and $\tilde{\Gamma}$ and Γ denoting the Dirac matrices associated with M and G/H respectively. If D is even, then the spinor representation of SO(1,D-1), defined by these matrices, is reducible, and we can define the chirality operator - in analogy to $\tilde{\Gamma}^5$:

$$\Gamma^{D+1} := (-i)^{(D-2)/2}\, \hat{\Gamma}^0 \hat{\Gamma}^1 \cdot \ldots \cdot \hat{\Gamma}^{D-1} \quad . \tag{3.1.12a}$$

We have

$$\hat{\Gamma}^{D+1} = \tilde{\Gamma}^5 \otimes \Gamma^{d+1} \quad , \tag{3.1.12b}$$

$$\Gamma^{d+1} = (-i)^{d/2}\, \Gamma^1 \ldots \Gamma^d \quad . \tag{3.1.12c}$$

Thus, for Weyl fermions on E, the eigenvalues of $\tilde{\Gamma}^5$ and Γ^{d+1} are correlated, giving the possibility to have a left-right asymmetry on $\tilde{M}$.

The choice of a (local) adapted orthonormal frame field $\{e_A\}$ on E is a section in $\hat{O}(E)$, which - together with section (1.2.16) - defines a (local) section in $\hat{Q} \longrightarrow E$. We denote the pull-back via this section of the G-invariant spinor field on $\hat{Q}$ under consideration by $\hat{\Psi}$, (spinor field on E). The action for this field has the following form [25]:

$$S_D = \int_E \left\{ -\tfrac{1}{2}\, \overline{\hat{\Psi}}\, \hat{\Gamma}^A\, e_A{}^M (\partial_M \hat{\Psi} - \omega_M \hat{\Psi} - \grave{\alpha}(\hat{A}_M)\hat{\Psi}) + \text{h.c.} \right\} \vartheta_E \,, \tag{3.1.13a}$$

with $\overline{\hat{\Psi}} = \hat{\Psi}^{\dagger}\hat{\Gamma}^{o}$, $e_A = e_A{}^M \partial_M$, and ω_M denoting the spin connection defined by

$$\omega_M := 1/2 \, \Gamma_{MAB} \, \Sigma^{AB} \quad . \tag{3.1.13b}$$

Here

$$\Gamma_{MAB} := \hat{\gamma}(\nabla_{e_A} e_B, \partial_M) \tag{3.1.13c}$$

are the Christoffel symbols corresponding to $\hat{\gamma}$ for the chosen frame field $\{e_A\}$, and

$$\Sigma^{AB} = 1/4 \, [\hat{\Gamma}^A, \hat{\Gamma}^B] \quad . \tag{3.1.13d}$$

Finally, α' denotes the representation of the Lie algebra k of K, induced by α , see (3.1.2).

For $\hat{\Psi}$ being G-invariant the Lagrangian (3.1.13a) is constant on G/H and, therefore, we can integrate over G/H. If we assume - for simplicity - that M is Minkowski space, then the reduced action has the following form [25] :

$$S_D = \mathrm{vol}(G/H) \int_{\tilde{M}} \Big\{ -\tfrac{1}{2} \overline{\Psi}_\alpha \tilde{\Gamma}^\mu (\partial_\mu \Psi_\alpha - \alpha'(A_\mu)\Psi_\alpha) -$$

$$- \tfrac{1}{2} \overline{\Psi}_\alpha (\tfrac{1}{8} f_{abc} \Gamma^a \Gamma^b \Gamma^c - \Gamma^a \alpha'(\phi_a))^\alpha{}_\beta \tilde{\Gamma}^5 \Psi^\beta + \text{h.c.} \Big\} d^4x, \tag{3.1.14}$$

with $f_{abc} := \gamma_{[1_G]}(u_a, [u_b, u_c]|_{\mathfrak{m}})$, $\{u_a\}$ being an orthonormal basis in $\mathfrak{m}$. The indices α, β denote the SO(d)-components of Ψ.

The reduced spinor field Ψ has to fulfill constraint equations (3.1.10), which infinitesimally look as follows:

$$\Delta'(\lambda'(h))\Psi + \alpha'(\tau'(h))\Psi = 0 \quad , \qquad , \tag{3.1.15a}$$

where the homomorphism $\Delta' \circ \lambda' : \mathfrak{h} \longrightarrow \mathrm{End}(\mathcal{R}_\Delta)$ is defined by

$$\Delta'(\lambda'(h)) = 1/2 \, \gamma_{[1_G]}([h, u_a], u_b) \Sigma^{ab} \quad , \tag{3.1.15b}$$

a,b = 1, ... , d.

Again, as in the case of bosonic matter fields, to obtain the explicit form of the reduced action, one has to solve the above constraints. We shall comment on this in section 3.3. There we shall also discuss several problems related to dimensional reduction of fermions.

We would, finally, like to underline that the requirement of G-invariance for fermions is quite restrictive; e.g. imposing rotation invariance for fermions on $M \times S^2$ completely eliminates them. Thus, it is interesting to study the procedure of harmonic expansion for fermions. A clear and mathematically consistent treatment of this subject can be found in [57].

3.2. The symmetry breaking scheme

In the framework of standard grand unified theories [1] and supersymmetric theories [2] one obtains a unification of all interactions, except the gravitational one, at an energy scale of 10^{14} - 10^{15} GeV. In supergravity, the unification scale turns out to be of the order of the Planck mass, $m_{Pl} \sim 10^{19}$ GeV. One of the main features of theories of this type is the existence of a gauge hierarchy: At energies of 10^{14} - 10^{15} Gev (or 10^{19} GeV respectively), the original gauge group is broken to $SU(3)_C \times SU(2)_W \times U(1)_Y$, and, subsequently, at energies $m_W \sim 100$ GeV it is broken further to $SU(3)_C \times U(1)_{em}$. Schematically, we shall write:

$$K \xrightarrow[10^{15}\text{ GeV}]{a_1} SU(3)_C \times SU(2)_W \times U(1)_Y \xrightarrow[10^{2}\text{ GeV}]{a_2} SU(3)_C \times U(1)_{em} \quad . \tag{3.2.1}$$

The most popular symmetry breaking mechanism is the Higgs mechanism. In theories of the above type the Higgs potential has to contain parameters, which "hierarchically" differ from each other. Therefore, one of the main problems in these theories is to explain the origin of scalar fields and potentials leading to (3.2.1). Moreover, in grand unified theories the masses and the number of fermion families are not predicted. As we shall see, there exists an interesting relation between the grand unification approach and the Kaluza-Klein approach to unification, discussed in this Review.

In the most general setting of a modern Kaluza-Klein theory [6,8] one considers a theory of gravity and (bosonic) matter, for example an Einstein-Yang-Mills system as discussed in section 2.3., on a multidimensional universe E. In a first step one looks for stable "ground state" solutions to the field equations exhibiting spontaneous compactification:

$$E = M \times G/H \quad . \tag{3.2.2}$$

Next one considers small fluctuations about the ground state values of the D-dimensional fields, linearizes the field equations with respect to these fluctuations and expands them harmonically on G/H. If one then integrates over G/H, one obtains an effective theory on 4-dimensional space time with an infinite tower of massive states with masses of the order of the Planck mass and a finite number of massless states. The observed particles ("low energy sector" of the theory) are supposed to correspond to those massless states. Finally, one hopes that these massless fields obtain their masses via quantum effects - a mechanism, which is far from being well-understood. Moreover, one should underline that the property of being massless or massive depends on the chosen ground state. Therefore, it is of special interest to investigate whether the ground state is stable or not, see also our discussion in section 2.3.

It turns out that a general analysis of the particle structure of the massless sector is rather impossible, (whereas it is possible to analyze the sector of symmetric fields in general terms - as we have shown in chapters 1 and 2.) However, it is guaranteed that one always has a massless graviton and massless Yang-Mills fields (provided the ground state admits Killing vectors - a condition trivially fulfilled for internal spaces of the form G/H). The further structure of the massless sector is highly model-dependent, see e.g. [6] .

It is interesting to apply the above-described procedure to pure Yang-Mills theories, see e.g. [13]. Then one obtains a symmetry breaking mechanism alternative to (3.2.1):

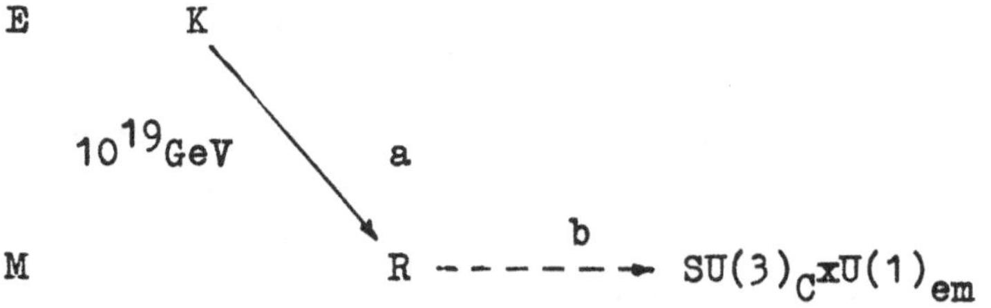

(3.2.3)

The interrupted line of step b in (3.2.3) means that the mechnism of breaking R to $SU(3)_C xU(1)_{em}$ is not specified in the above approach, it can happen, for example, dynamically.

If we restrict ourselves to the sector of symmetric fields (and models giving after dimensional reduction a Higgs potential, see chapter 1), then we have the following mechanism:

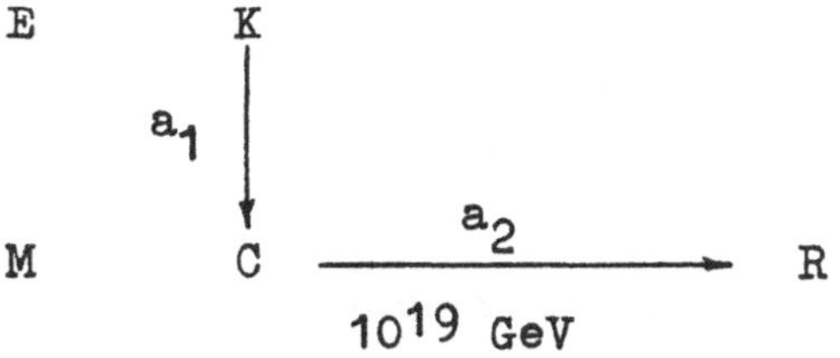

. (3.2.4)

Step a_1 corresponds to the geometrical symmetry breaking from K to C, the centralizer of $\tau(H)$ in K, and step a_2 corresponds to spontaneous symmetry breaking via the Higgs potential obtained in the dimensional reduction procedure.

The above ideas have been also applied to supersymmetric Yang-Mills theories, see e.g. [54] for an application to D = 10 minimal super-Yang-Mills with $K = E_8$. It is well known that one of the problems in supersymmetric theories is to find a supersymmetry breaking mechanism, which guarantees that the supersymmetric partners of physical particles obtain big masses. It was shown for the model considered in [54] that in step a_1 of (3.2.4) not only the gauge group K, but also the supersymmetry is broken; and that after step 2 one can obtain an interesting low energy sector. It was also shown that the fact, whether N = 1 supersymmetry is broken via dimensional reduction or not, depends only on the properties of the internal space G/H, see e.g. [53] .

We shall call models for which the multidimensional universe $E = M \times G/H$ is obtained via solving the equations of spontaneous compactification, and for which, consequently the size of G/H is of the order of the Planck length, models of type I.

There is another type of models, called further on models of type II, frequently discussed in the literature. In these models one does not insist that $E = M \times G/H$ should be obtained by the spontaneous compactification mechanism. One just chooses multidimensional universes $E = M \times G/H$ (and gauge groups K), such that one is able to construct models using the CSDR-scheme discussed in detail in chapter 1. In this type of models the size of the internal space G/H strongly differs from m_{Pl}^{-1} , being either of the (inverse) order of the grand unification scale, $m \sim 10^{15}$ GeV, or of the electroweak scale, $m \sim 10^2$ GeV. The symmetry breaking mechanism in this approach is given by (3.2.4), with the only difference that the energy scale in step a_2 is different from m_{Pl} .

Using the model building scheme II, one can construct the bosonic sector of the Weinberg-Salam model [25,60,49,53,79] or the unification model of strong and electroweak interactions [53,63] ; the symmetry breaking scheme for these cases is:

$$\begin{array}{lll} E & K & \\ & \searrow & C = SU(2)_W \times U(1)_Y \xrightarrow{\;a_2\;} R = U(1)_{em} \\ M & a_1 \downarrow & \\ & & C = SU(3)_C \times SU(2)_W \times U(1)_Y \xrightarrow[10^2\,\mathrm{GeV}]{} R = SU(3)_C \times U(1)_{em} \, . \end{array} \tag{3.2.5}$$

Another interesting application is to obtain a grand unified model after the first (geometric) symmetry breaking, that means C is one of the grand unified groups or contains $SU(3)_C \times SU(2)_W \times U(1)_Y$ as a subgroup, see [52,61-64, 112] . We shall comment on this concept in section 3.4. Here we only mention that grand unified models obtained in this way do not suffer from a high number of free parameters. Therefore, one can hope to obtain more predictive models than in the standard approach. (However, almost all activities in this direction till now are in the very special class of $G \subset K$.)

3.3. Comments on the fermionic sector

We have already discussed in section 3.1. how to deal with fermions within the dimensional reduction method. For a discussion of model building including fermions we refer to [25,20,52-54,56,59,62,63,65,66,117] . Let us adopt the notation introduced in section 3.1. and let us assume that E is even-dimensional. Then the chirality operator $\hat{\Gamma}^{D+1}$, see (3.1.12a), acts on $\mathcal{R}_\Delta$ reducibly, and we have the decomposition

$$\Delta = \Delta^+ \oplus \Delta^- \tag{3.3.1}$$

into irreducible representations corresponding to chirality ± 1. If we reduce SO(1,D-1) to SO(1,3) x SO(d) , as discussed in section 3.1., we obtain the following decompositions of the corresponding restrictions of Δ^+ and Δ^- :

$$\Delta^+ \restriction_{SO(1,3)xSO(d)} = (\Delta_4^+, \Delta_d^+) \oplus (\Delta_4^-, \Delta_d^-) \quad , \tag{3.3.2a}$$

$$\Delta^- \restriction_{SO(1,3)xSO(d)} = (\Delta_4^+, \Delta_d^-) \oplus (\Delta_4^-, \Delta_d^+) \quad . \tag{3.3.2b}$$

Let us now consider a spinor field $\hat{\Psi}$ on E of given chirality, say +1, transforming under the representation $\mathcal{D} = (\Delta^+, \alpha)$ of SO(1,D-1)xK . As we have shown in section 3.1., a G-symmetric spinor field $\hat{\Psi}$ on E is equivalent to a spinor field Ψ on M, fulfilling constriants (3.1.15). To find the representation, say $\bar{\mathcal{D}}$, under which Ψ transforms, explicitly, one has to solve (3.1.15a). Using (3.3.2a) we have

$$\mathcal{D} \restriction_{SO(1,3)xSO(d)xK} = (\Delta_4^+, \Delta_d^+, \alpha) \oplus (\Delta_4^-, \Delta_d^-, \alpha) \ . \tag{3.3.3}$$

Let us - for simplicity - assume that there are no U(1)-factors in H. Denoting the decompositions into irreducible components of $\Delta_d^+ \restriction_{\lambda(H)}$ and $\alpha \restriction_{\tau(H)xC}$ by

$$\Delta_d^{\pm}\upharpoonright\lambda(H) = \bigoplus_j \varepsilon_j^{\pm} \quad , \tag{3.3.4a}$$

$$\alpha\upharpoonright\tau(H)\times C = \bigoplus_k (\alpha_k, \chi_k) \quad , \tag{3.3.4b}$$

we can restrict $\mathcal{D}$ further:

$$\mathcal{D}\upharpoonright_{SO(1,3)\times\beta(H)\times C}$$

$$= \Big\{ \bigoplus_{j,k} (\Delta_4^{+}, \varepsilon_j^{+}\otimes\alpha_k, \chi_k) \Big\} \oplus \Big\{ \bigoplus_{j,k} (\Delta_4^{-}, \varepsilon_j^{-}\otimes\alpha_k, \chi_k) \Big\} \ . \tag{3.3.4c}$$

But (3.1.15a) means that Ψ is in the trivial representation of $\beta(H)$, see also (3.1.10). From Schur's Lemma we have that the trivial representation occurs only for pairs (j,k), for which $\varepsilon_j^{\pm}$ is contragradient to α_k, $\varepsilon_j^{\pm} = \alpha_k^{*}$. Therefore, we get:

$$\bar{\mathcal{D}} = \Big\{ \bigoplus_k (\Delta_4^{+}, \chi_{k+}) \Big\} \oplus \Big\{ \bigoplus_k (\Delta_4^{-}, \chi_{k-}) \Big\} \ . \tag{3.3.4d}$$

The summation index $k^{+}(k^{-})$ is defined as follows: With every pair $\varepsilon_j^{+}, \alpha_k$ ($\varepsilon_j^{-}, \alpha_k$), such that $\varepsilon_j^{+} = \alpha_k^{*}$ ($\varepsilon_j^{-} = \alpha_k^{*}$), we associate a representation $\chi_{k+} := \chi_k$ ($\chi_{k-} := \chi_k$). If

$$\chi^{(L)} = \bigoplus_{k^+} \chi_{k+} \neq \chi^{(R)} = \bigoplus_{k^-} \chi_{k-} \quad , \tag{3.3.5}$$

then the reduced theory is chiral. It was shown in [25] that a theory with symmetric Weyl fermions on E reduces to a chiral model on M only if rank(G) = rank(H) , H has complex representations and α is not self-conjugate. The second and the third point can be easily deduced from the above considerations, but the first point follows from non-trivial index theorem arguments due to Bott [113]. Some remarks on the connection between chiral asymmetry and the G-index [113] of the Dirac operator on G/H can be also found in [25]. There it was argued that the difference in the number of right- and left-handed fermions in 4 dimensions should be given by the singlet piece of the G-index. Moreover, an explicit example of a chiral theory obtained via dimensional reduction was discussed. For further attempts to construct theories in 4 dimensions containing chiral fermions see [117], and further references therein.

It is of interest to ask, whether chiral fermions can survive the spontaneous symmetry breaking from C to R, step a_2 in (3.2.4). This question was considered in [59] for the case $H \subset G \subset K$, and the answer is negative. This could be, from the phenomenological point of view, allowed only for the case when $R = SU(3)_C x U(1)_{em}$. But, as shown in [63,66], for $H \subset G \subset K$ and $R = SU(3)_C x U(1)_{em}$ it is impossible to obtain a fermionic sector with correct quantum numbers within the CSDR-scheme. It follows from these considerations that one should rather depart from the case $H \subset G \subset K$ and investigate models, for which τ cannot be continued to a homomorphism from G to K, see [50,51,63,64,75]. However, for this case, till now there are no systematic results concerning the fermionic sector after spontaneous symmetry breaking.

Finally, let us comment on the case $D = 4n+2$. Then Weyl and Mayorana conditions on the spinor fields can be implemented simultaneously, and one can obtain chiral fermions after dimensional reduction even if the representation α is self-conjugate [53]. (Vector-like representations often occur in supersymmetric theories.)

3.4. Realistic models

In this section we discuss a few models of type II, obtained via the CSDR-method. In several papers [60, 49,79] the bosonic sector of the Weinberg-Salam model has been constructed. Some attempts to obtain the Weinberg-Salam model including the fermionic sector have been made, too [25,53]. But, till now, no satisfying solution to this problem has been found. In the first paper on this subject [60] the multidimensional universe was of the form $E = M \times S^2$, $S^2 \cong SU(2)/U(1)$, and the gauge group K was either SU(3) or SO(5) or G_2. It was shown that in these cases one obtaines $C = SU(2)_W x U(1)_Y$, a Higgs doublet in the fundamental representation and a generic Higgs potential giving spontaneous symmetry breaking to $U(1)_{em}$. The results of [60] have been essentially generalized in [49,79] . In these papers the case of symmetric spaces G/H, with G and H being arbitrary classical Lie groups, arbitrary simple classical Lie groups K and regular embeddings $H \subset G$ and $\tau(H) \subset K$ has been considered. An exhaustive list of models fulfilling these asumptions, and giving the bosonic sector of the Weinberg-Salam model can be found in Table 1. The physical parameters of this model can be - after an appropriate rescaling of the fields, leading to the canonical form of the reduced action - calculated in terms of the gauge coupling constant g of the multidimensional theory, the parameter m characterizing the size of G/H and the dimension d of G/H. In particular, one can calculate the masses m_W and m_Z of the intermediate vector bosons and the mass m_H of the Higgs boson after spontaneous symmetry breaking, see Table 1. Thus, one gets a prediction for the mass of the Higgs bosons; typically, we obtain that it is of the order of m_Z , very often we have $m_H = m_Z$. In Table 1 we also listed the value of $\sin^2\Theta_W$, with Θ_W denoting the Weinberg angle,

$$\cos^2\Theta_W = m_W^2/m_Z^2 .$$

We note that $\sin^2\Theta_W$ depends only on d. Thus, the dimension of the internal space can be fixed, demanding $\sin^2\Theta_W$ to be as near to the experimental value as possible,

$$(\sin^2\Theta_W)_{exp} = 0{,}233 \pm 0{,}009 , \quad \text{see [118]} .$$

Taking into account this requirement, we see from Table 1 that theory 2 (for d = 6 and d = 8) and theories 3 and 4 are good candidates for realistic models, giving $\sin^2\Theta_W = 0{,}25$ or $\sin^2\Theta_W = 0{,}2$. One can try to explain the difference from the experimental value by the help of quantum corrections, see [79]. Moreover, one can adjust the parameters g and m to give the correct experimental values of, say, the fine structure constant and m_Z. This fixes the size of the internal space. As already mentioned, one typically obtains $m \sim 10^2$ GeV, corresponding to a size of G/H of the order of 10^{-16} cm.

As an illuminating example with two dimensional parameters let us consider $E = M \times SO(10)/SU(5)$ and

$K = E_8$, see [52] . One obtains $C = SU(5)$ and, since τ can be extended to a homomorphism from $G \cong SO(10)$ to $K \cong E_8$, one has $R = SU(4)$. The decompositions of $ad_{\mathfrak{h}}(\mathfrak{m})$ and $ad_{\mathfrak{k}}|_{\tau'(\mathfrak{h}) \oplus \mathfrak{c}}$, see (1.3.5), into irreducible components of $\mathfrak{h} = su(5)$ and $\tau'(\mathfrak{h}) \oplus \mathfrak{c} = su(5) \oplus su(5)$, respectively, are given by:

$$ad_{\mathfrak{h}}(\mathfrak{m}) = \underline{1} \oplus \underline{10} \oplus \underline{10}^* \quad , \tag{3.4.1a}$$

$$ad_{\mathfrak{k}}|_{\tau'(\mathfrak{h}) \oplus \mathfrak{c}}$$
$$= [\underline{5},\underline{10}^*] \oplus [\underline{5}^*,\underline{10}] \oplus [\underline{10},\underline{5}] \oplus [\underline{10}^*,\underline{5}^*] \oplus [\underline{24},\underline{1}] \oplus [\underline{1},\underline{24}] \quad . \tag{3.4.1b}$$

Thus, we get two irreducible scalar field multiplets of $C = SU(5)$, one in the fundamental ($\underline{5}$) and one in the adjoint representation ($\underline{24}$). This means that we have obtained the scalar fields of the minimal SU(5) grand unified model [1]. The number of vector bosons getting a mass after symmetry breaking from C to R is dimC-dimR = 9 . The mass of one of those bosons, corresponding to the generator of the U(1)-subgroup of SU(4)xU(1) SU(5), is equal to $m_A = \sqrt{8}\, m_2$. The masses of the remaining bosons are given by

$$m_B = 2\, m_2 \cdot \sqrt{1 + (5 m_1)^2/(8 m_2)^2} \quad .$$

In these formulae m_1 and m_2 are parameters, characterizing the SO(10)-invariant metric on G/H, with m_1 corresponding to the invariant subspace of the $\underline{10}$ - representation of $\mathfrak{h}$. If $m_1 \gg m_2$, one could hope to have a gauge hierarchy with intermediate gauge group $SU(4) \times U(1)$. Unfortunately, in the sector of scalar fields one has only bosons with masses of the order m_2 and $m_2^2/m_1 \ll m_2$; there are no Higgs bosons with masses of the order of m_B, indicating that the standard hierarchical symmetry breaking mechanism does not occur.

For a discussion of another example with two dimensional parameters we refer to [51]. In this paper also the case of non-injective homomorphisms $\tau : H \longrightarrow K$ has been discussed.

In [53,59,62,65-67] there have been made attempts to construct grand unified models including fermions.

The main point of these papers was to analyze, which fields are left after dimensional reduction. Such questions as the explicit form of the reduced action or the form of the gauge group R after spontaneous symmetry breaking were not analyzed systematically. As a typical example [67] , let us consider the minimal super Yang-Mills theory with gauge group E_8 on the 10-dimensional universe E = M x Sp(2)/SU(2)xU(1) , with an embedding of SU(2)xU(1) into Sp(2) such that Sp(2)/SU(2)xU(1) is not symmetric. For the reduced gauge group one obtains C = SU(3)xSU(5)xU(1
The decompositions of $\mathrm{ad}\,\mathfrak{g}\restriction\mathfrak{h}$ and $\mathrm{ad}\,\mathfrak{k}\restriction\tau'(\mathfrak{h})\oplus\mathfrak{l}$, see (1.3.5), are given by:

$$\mathrm{ad}\,\mathfrak{g}\restriction\mathfrak{h} = \underline{3}(0)\oplus\underline{1}(0)\oplus\underline{1}(2)\oplus\underline{1}(-2)\oplus\underline{2}(1)\oplus\underline{2}^{*}(-1) \quad ,(3.4.2a)$$

$$\begin{aligned}&\mathrm{ad}\,\mathfrak{k}\restriction\tau'(\mathfrak{h})\oplus\mathfrak{l}\\ &\quad= [\underline{1},\underline{1},\underline{5}](6)\oplus[\underline{1},\underline{3}^{*},\underline{5}](-4)\oplus[\underline{2},\underline{3},\underline{5}](1)\oplus[\underline{2},\underline{1},\underline{10}](3)\\ &\qquad+[\underline{1},\underline{3},\underline{10}^{*}](-2)\oplus[\underline{2},\underline{3},\underline{1}](-5)\oplus \text{h.c.}\oplus\mathrm{ad}\,\tau'(\mathfrak{h})\oplus\mathrm{ad}\,\mathfrak{l}' ,\end{aligned}\tag{3.4.2b}$$

where in quadratic brackets we have specified the type of the representations of $\mathfrak{h}'$ = su(2) and $\mathfrak{l}'$ = su(3) ⊕ su(5) - the non-Abelian parts of $\mathfrak{h}$ and $\mathfrak{l}$. (The above decompositions can be found, using e.g. the tables given in [88].) For $\mathrm{ad}\,\tau'(\mathfrak{h})$ and $\mathrm{ad}\,\mathfrak{l}'$ we have:

$$\mathrm{ad}\,\tau'(\mathfrak{h}) = [\underline{3},\underline{1},\underline{1}](0)\oplus[\underline{1},\underline{1},\underline{1}](0) \quad ,$$

$$\mathrm{ad}\,\mathfrak{l}' = [\underline{1},\underline{8},\underline{1}](0)\oplus[\underline{1},\underline{1},\underline{24}](0) \quad .$$

Using Schurs Lemma we find that the scalar fields are in the representation $[\underline{3},\underline{10}^{*}](-2)\oplus[\underline{3},\underline{5}](1)$ of C = SU(3)xSU(5). Now, let us find the fermionic multiplets. The Weyl-Mayorana spinors transform under the adjoint representation $\alpha = \mathrm{ad}\,\mathfrak{k} \equiv 248$ of $K=E_8$. The decomposition (3.3.4a) has the form

$$\Delta_6^{+}\restriction\lambda(H) = \underline{1}(2)\oplus\underline{1}(0)\oplus\underline{2}(-1) \quad , \tag{3.4.3}$$

decomposition (3.3.4b) coincides with (3.4.2b). Therefore, the left-handed fermions transform under the chiral representation

$$[\underline{3},\underline{10}^*](-2) \oplus [\underline{3},\underline{5}](1) \oplus [\underline{1},\underline{1}](0) \oplus [\underline{8},\underline{1}](0) \oplus [\underline{1},\underline{24}](0)$$

of the reduced gauge group C. We see that we have an N = 1 supersymmetry in the reduced theory. In particular, the fermionic sector contains the sector of the minimal SU(5) model with three fermion families, provided we identify $SU(3) \subset C$ with the flavour symmetry. Unfortunately, the scalar field sector contains only fields responsible for the electroweak symmetry breaking - and no fields in the adjoint representation, giving the reduction from SU(5) to SU(3)xSU(2)xU(1). Other models were also considered in [67], but till now there has not been found any model giving a 4-dimensional theory, which would be completely satisfactory from the phenomenological point of view.

May be, one should investigate models including Higgs fields on the level of the multidimensional theory, e.g. N = 4 super-Yang-Mills with gauge group E_8 . On the other hand, recently there have been proposed other geometrical symmetry breaking mechanisms, e.g. using Wilson loops [114,115] , which possibly could be included into the coset space dimensional reduction scheme. However, all these points need further investigations.

References

[1] - M.Gell-Mann,P.Ramond,R.Slansky, Rev.Mod.Phys. (1978), Vol.50, No.4, 721

- S.G.Matinjan, Usp.Fiz.Nauk (1980),Vol.130,No.1, 3
- P.Langacker, Phys.Rep.C (1981),Vol.72,No.4, 185
- P.Ramond, Ann.Rev.Nucl.Part.Sc. (1983),Vol.33, 31

[2] - V.I.Ogievetsky,L.Mezinchesky, Usp.Fiz.Nauk (1975),Vol.117, No.4, 637

- P.Fayet,S.Ferrara, Phys.Rep.C (1977),Vol.32,No.5, 249
- A.A.Slavnov, Usp.Fiz.Nauk (1978), Vol.124,No.3, 487

[3] - P.van Nieuwenhuizen, Phys.Rep.C (1981),Vol.68,No.4, 189

- J.Wess,J.Bagger, Supersymmetry and supergravity, Princeton University Press, Princeton 1983
- P.West, Introduction to Supersymmetry and Supergravity, World Scient.Publ. 1986

[4] - T.Kaluza, Sitzungsber.Preuss.Akad.Wiss., Berlin (1921), Math.Phys. K1, 966

- C.Klein, Z.Phys. (1926), Vol.37, 895

[5] - P.Bergmann,A.Einstein, Ann.Math. (1938),Vol.39, 683

- V.Fock, Z.Phys. (1926),Vol.38, 242 and Vol.39, 226

[6] - M.J.Duff,B.Nilsson,C.Pope, Phys.Rep.C (1986),Vol.130,No.1/2,1

- M.J.Duff,"Modern Kaluza-Klein Theories", Kaluza-Klein Workshop, Chalk River 1983, prepr.Imp.Coll. TP 83-84/45

[7] - J.Schwarz, Phys.Rep.C (1982),Vol.89,No.3, 223

[8] - A.Salam,J.Strathdee, Ann.Phys. (1982),Vol.141,No.2, 316

[9] - W.Mecklenburg, Fortschr.d.Phys. (1984),Vol.32,No.5, 207

[10] - I.Ya.Arefeva,I.V.Volovich, Usp.Fiz.Nauk (1985),Vol.146, No.4, 655

[11] - D.P.Sorokin,V.I.Tkach, ECHAYA (1987),Vol.18,No.5, 1035

[12] - E.Cremmer,J.Scherk, Nucl.Phys.B (1977),Vol.118,No.1/2, 61

[13] - Z.Horvath,L.Palla,E.Cremmer,J.Scherk,Nucl.Phys.B (1977), Vol.127,No.1, 57

[14] - I.F.Luciani, Nucl.Phys.B (1978),Vol.135,No.1, 111

[15] - D.V.Volkov,V.I.Tkach, Pisma ZETF (1980),Vol.32,No.2, 681

[16] - D.V.Volkov,V.I.Tkach, Teor.Mat.Fiz. (1982),Vol.51,No.2, 171

[17] - D.V.Volkov,D.P.Sorokin,V.I.Tkach, Pisma ZETF (1983),Vol.38, No.8, 397 and Teor.Mat.Fiz. (1984),Vol.61, 241

[18] - S.Randjbar-Daemi,R.Percacci, Phys.Lett.B (1982),Vol.117, No.1/2, 41

[19] - S.Randjbar-Daemi,A.Salam,J.Strathdee, Nucl.Phys.B (1983), Vol.214,No.3, 491 and Phys.Lett.B (1983),Vol.124,No.5, 345

[20] - A.N.Schellekens, Nucl.Phys.B (1984),Vol.248,No.3, 706 and Phys.Lett.B (1984),Vol.143,No.1, 121

[21] - K.Pilch,A.N.Schellekens, Nucl.Phys.B (1985),Vol.256,No.1, 109

[22] - P.Forgacs,Z.Horvath,L.Palla, Z.Phys.C (1986),Vol.30,No.2, 261

[23] - P.G.O.Freund,M.A.Rubin, Phys.Lett.B (1980),Vol.97,No.2, 233

[24] - Z.Horvath,L.Palla, Nucl.Phys.B (1978),Vol.142,No.3, 327

[25] - N.S.Manton, Nucl.Phys.B (1981),Vol.193,No.2, 502

[26] - S.Randjbar-Daemi,A.Salam,J.Strathdee, Phys.Lett.B (1983), Vol.132,No.1, 56

- P.H.Frampton,K.Yamamoto, Nucl.Phys.B (1985),Vol.254,No.2, 349

[27] - E.Witten, Nucl.Phys.B (1981),Vol.186,No.3, 412

[28] - P.Candelas,G.Horowitz,A.Strominger,E.Witten, Nucl.Phys.B (1985),Vol.258,No.1, 46

- A.Strominger,E.Witten, Comm.Math.Phys. (1985),Vol.101,No.3, 341

- L.Dixon,J.Harvey,C.Vafa,E.Witten, Nucl.Phys.B (1986), Vol.261, 678 and Vol.274, 285

[29] - T.R.Govindarajan,A.S.Joshipura,S.D.Rindani,U.Sarkar,"Coset Spaces as alternatives to Calabi-Yau spaces in the presence of gaugino condensation", prepr.ICTP 10/86/170 (1986)

[30] - J.A.Wolf, Spaces of Constant Curvature, McGraw-Hill, New York 1967

- S.Helgason,Differential Geometry and Symmetric Spaces, Acad.Press, New York, 1962

- S.Kobayashi,K.Nomizu, Foundations of Differential Geometry, Vol.2, Wiley (Interscience), New York 1969

[31] - J.A.Wolf, Acta Math. (1968),Vol.120, 59

[32] - I.Ya.Arefeva,I.V.Volovich, Usp.Fiz.Nauk (1985),Vol.146, No.4, 655 and Phys.Lett.B (1985),Vol.164,No.4/5/6, 287

[33] - C.Wetterich, Nucl.Phys.B (1984),Vol.244,No.2, 359 and Nucl.Phys.B (1985),Vol.252,No.1, 309

[34] - V.A.Rubakov,M.B.Shaposhnikov, Phys.Lett.B (1983), Vol.125, No.2, 139

[35] - D.P.Sorokin,V.I.Tkach, ECHAYA (1987),Vol.18,No.5, 1035

[36] - M.J.Duff,C.N.Pope, Nucl.Phys.B (1985),Vol.225,No.2, 355

[37] - L.Palla, Z.Phys.C (1984),Vol.24,No.2, 195

[38] - E.Witten, Phys.Rev.Lett. (1977),Vol.38,No.3, 121

[39] - A.S.Schwarz, Comm.Math.Phys. (1977),Vol.56,No.1, 79
- V.A.Romanov,A.S.Schwarz,Yu.S.Tyupkin, Nucl.Phys.B (1977), Vol.130,No.2, 209

[40] - A.Trautman, Bull.Acad.Polon.Sci., ser.sci.phys.et astron. (1979),Vol.XXVII,No.1, 7 and Lectures at XX.Universitätswochen f.Kernphysik (Schladming), Acta Phys.Austr. (1981) Suppl.23, 401

[41] - P.Forgacs,N.S.Manton, Comm.Math.Phys. (1980),Vol.72,No.1, 15

[42] - J.Harnad,S.Shnider,L.Vinet, J.Math.Phys. (1980),Vol.21, No.12, 2719

[43] - J.Harnad,S.Shnider,J.Tafel, Lett.Math.Phys. (1980),Vol.4, No.2, 107

[44] - G.Rudolph,I.P.Volobujev, in Geom.Meth.in Phys.Brno (1984), Proc.Conf.on Diff.Geom.and its Appl., Nové Město 1983, 239

[45] - A.Jadczyk,K.Pilch, Lett.Math.Phys. (1984),Vol.8,No.2, 97

[46] - I.P.Volobujev,G.Rudolph, Teor.Mat.Fiz. (1985),Vol.62,No.3, 388

[47] - L.B.Hudson,R.Kantowski, J.Math.Phys. (1984),Vol.25,No.10, 3094
- L.Nikolova,V.Rizov, Lett.Math.Phys. (1985),Vol.10, 315 and J.Math.Phys. (1986),Vol.27,No.1, 132
- P.Nikolov, J.Math.Phys. (1987),Vol.28,No.10, 2354
- M.Chaichian,A.P.Demichev,N.F.Nelipa,A.Yu.Rodionov, Nucl. Phys.B (1987),Vol.279, 452

[48] - I.P.Volobujev,Yu.A.Kubyshin, Teor.Mat.Fiz. (1986),Vol.68, No.2, 225

[49] - I.P.Volobujev,Yu.A.Kubyshin, Teor.Mat.Fiz. (1986),Vol.68, No.3, 368

[50] - Yu.A.Kubyshin,J.M.Mourao,I.P.Volobujev,"Scalar fields in the dimensional reduction scheme for symmetric spaces", preprint Lisbon IFM-15/87 (1987)

[51] - G.Rudolph,I.P.Volobujev, Nucl.Phys.B (1989),Vol.313, 95

[52] - G.Chapline,N.S.Manton, Nucl.Phys.B (1981),Vol.184,No.3, 39

[53] - G.Chapline,R.Slansky, Nucl.Phys.B (1982),Vol.209,No.2,461

[54] - D.Olive,P.West, Nucl.Phys.B (1983),Vol.217,No.1, 248

[55] - C.Wetterich, Nucl.Phys.B (1984),Vol.242,No.2, 473

[56] - G.Chapline,B.Grossmann, Phys.Lett.B (1984),Vol.135,No.1, 109

[57] - R.Coquereaux,A.Jadczyk, Class.Quant.Grav. (1986),Vol.3, No.1, 29

[58] - N.S.Manton, Ann.Phys. (1986),Vol.167,No.2, 328

[59] - K.J.Barnes,R.C.King,M.Surridge, Nucl.Phys.B (1987),Vol.281 No.1, 253

[60] - N.S.Manton, Nucl.Phys.B (1979),Vol.158,No.1, 141

[61] - P.Forgacs,G.Zoupanos, Phys.Lett.B (1984),Vol.148,No.1/2/3, 99

[62] - K.J.Barnes,M.Surridge, Z.Phys.C (1986),Vol.33,No.1, 89

[63] - F.A.Bais,K.J.Barnes,P.Forgacs,G.Zoupanos, Nucl.Phys.B (1986), Vol.263,No.3/4, 557

[64] - K.Farakos,G.Koutsoumbas,M.Surridge,G.Zoupanos, Nucl.Phys.B (1987),Vol.291,No.1, 128 and Phys.Lett.B (1987),Vol.191, No.1/2, 135

[65] - D.Lüst,G.Zoupanos, Phys.Lett.B (1985),Vol.165,No.4/5/6, 309

[66] - K.J.Barnes,P.Forgacs,M.Surridge,G.Zoupanos, Z.Phys.C (1987), Vol.33,No.3, 427

[67] - N.N.Kozimirov,I.I.Tkachev, Z.Phys.C (1987),Vol.36,No.1, 83

[68] - B.S.de Witt, in Relativity,Groups and Topology, Gordon and Breach, NY 1964

- J.Rayski, Acta Phys.Polon. (1965),Vol.27,No.6, 947 and Vol.28,No.1, 87

- R.Kerner, Ann.Inst.H.Poinc.A (1968),Vol.9,No.2, 143

[69] - A.Trautman, Rep.Math.Phys. (1970),Vol.1,No.1, 29

[70] - Y.M.Cho, J.Math.Phys. (1975),Vol.16,No.10, 2029

- Y.M.Cho,P.G.O.Freund, Phys.Rev.D (1975),Vol.12,No.6, 1711

- Y.M.Cho,P.S.Jang, Phys.Rev.D (1975),Vol.12,No.12, 3789

[71] - R.Coquereaux,A.Jadczyk, Comm.Math.Phys. (1983),Vol.90, No.1, 79

[72] - P.Forgacs,Z.Horvath,L.Palla, Phys.Lett.B (1984),Vol.147, No.4/5, 311

[73] - I.P.Volobujev,Yu.A.Kubyshin, Pisma ZETF (1987),Vol.45,No.10, 455 and Teor.Mat.Fiz. (1988),Vol.75,No.2, 255

[74] - Yu.A.Kubyshin,J.M.Mourao,I.P.Volobujev, Phys.Lett.B (1988), Vol.203,No.4, 349

[75] - M.Surridge, Z.Phys.C (1987),Vol.37,No.1, 77

[76] - T.Appelquist,A.Chodos, Phys.Rev.D (1983),Vol.28,No.4, 772

[77] - T.Appelquist,A.Chodos, Phys.Rev.Lett. (1983),Vol.50,No.3, 141

[78] - M.J.Duff,D.J.Toms,"Divergencies and anomalies in Kaluza-Klein theories", CERN-prepr. TH 3248 (1982)

[79] - I.P.Volobujev,Yu.A.Kubyshin, in "Kwarki 86" Tbilissi 1986, Moscow 1987, 165 and in "Trudy IX.seminara po fizike wysokich energii i teorii pol`a" Protvino 1986, Moscow 1987, 153

[80] - T.Hübsch,"Calabi-Yau manifolds - Motivations and constructions", University of Maryland Report PP86-149

[81] - T.Eguchi,P.B.Gilkey,A.J.Hanson, Phys.Rep.C (1980),Vol.66, No.6, 213

- W.Drechsler,M.E.Mayer, Lect.Notes in Phys. 67, Springer-Verlag, Berlin 1977
- M.Daniel,C.M.Viallet, Rev.Mod.Phys. (1980),Vol.52, 175
- M.F.Atiyah,"Geometry of Yang-Mills fields", Acad.Naz.dei Lincei, Pisa 1979

[82] - S.Kobayashi,K.Nomizu, Foundations of Differential Geometry, Vol.2, Wiley (Interscience), New York 1963

[83] - G.E.Bredon, Introduction to compact transformation groups, Acad.Press, New York - London 1972

[84] - M.Henneaux, J.Math.Phys. (1982),Vol.23,No.5, 830
- A.Jadczyk, J.Geom.Phys. (1984),Vol.1, 97

[85] - G.Rudolph, Lett.Math.Phys. (1987),Vol.14, 133 and in Lect. Notes in Phys. 313 (1987), 485

[86] - E.Cremmer,B.Julia, Nucl.Phys.B (1979),Vol.159,No.1/2, 141

[87] - E.B.Dynkin, Mat.zbornik (1952),Vol.30,No.2, 349

[88] - R.Slansky, Phys.Rep.C (1981),Vol.79,No.1, 1

[89] - W.McKay,J.Patera, Tables of dimensions,indices and branchin rules for representations of simple Lie algebras, New York Dekker 1981

[90] - A.A.Kirillov, Elementy teorii predstavlenij, Nauka, Moskva 1978

[91] - N.Jacobson, Lie algebras, Interscience Publ., New York 1962

[92] - M.Goto,F.Grosshans, Semisimple Lie algebras, Lect.Notes in Pure and Appl.Math., Vol.38, Marcel Dekker,INC., New York and Basel 1978

[93] - E.B.Dynkin, Trudy Mosk.Mat.obschestva (1952),Vol.1, 39

[94] - A.Trautman, Ist.Naz.di Alta Math., Simposia Math. (1973), Vol.XII, 139

[95] - M.J.Duff,C.A.Orzalezi, Phys.LettB (1983),Vol.122, 37
- Y.S.Wu,A.Zee, J.Math.Phys. (1984),Vol.25, 2696
- Y.S.Wu, Phys.Rev.D (1984),Vol.29, 2796
- I.J.Muzinich, J.Math.Phys. (1985),Vol.26,No.8, 1942

[96] - D.E.Neville, Phys.Rev.D (1986),Vol.33,No.2, 363

[97] - C.A.Orzalezi,M.Pauri, Phys.Lett.B (1981), Vol.107, 186
- C.A.Orzalezi,G.Venturi, Phys.Lett.B (1984),Vol.139, 357

[98] - B.McInnes, J.Math.Phys. (1986),Vol.27,No.8, 2029

[99] - G.Rudolph, J.Geom.Phys. (1987),Vol.4,No.1, 39 and Proc. of the XX.Int.Symp. Ahrenshoop on the Theory of Elem.Part. 198

[100] - I.V.Volovich,M.O.Katana'ev, Teor.Mat.Fiz. (1986),Vol.66, No.1, 79

[101] - R.Coquereaux, Acta Phys.Pol.B(1984),Vol.15,No.9, 821

[102] - F.Müller-Hoissen, Class.Quant.Grav. (1987),Vol.4, L143

[103] - B.Bantay, Mod.Phys.Lett.A (1987),Vol.2,No.1, 57

[104] - F.Müller-Hoissen, Phys.Lett.B (1985),Vol.163,No.1/2/3/4, 106

[105] - A.D.Popov, Teor.Mat.Fiz. (1987),Vol.71,No.1, 67 and Yadernaja Fiz. (1987),Vol.46,No.1, 289

[106] - D.V.Volkov,D.P.Sorokin,V.I.Tkach, Pisma ZETF (1984),Vol.40, No.8, 356 and Yadernaja Fiz. (1984),Vol.39,No.5, 1306,(1985), Vol.41,No.5, 1373, (1986),Vol.43,No.1, 222, and Phys.Lett.B (1985),Vol.161,No.4/5/6, 301

[107] - B.E.W.Nilsson,C.N.Pope, Class.Quant.Grav. (1984),Vol.1, No.5, 499

- B.Dolan, Phys.Lett.B(1984),Vol.140,No.5/6, 304

[108] - S.Coleman, in Proc.of the Int.School of Subnucl.Phys. 1975, Part A, Ed. by Zichichi, London Plenum Press 1977, 297

[109] - S.Weinberg,P.Candelas, Nucl.Phys.B (1984),Vol.223, 433

- A.Chodos,E.Myers, Ann.Phys. (1984),Vol.156, 412

- Nguyen Van Hieu, Fortschr.d.Phys. (1986),Vol.34, 441

- M.H.Sarmadi, Nucl.Phys.B (1986),Vol.263, 187

[110] - J.Milnor, L'Enseignement Math. (1963),Vol.9, 198

- K.Bichteler, J.Math.Phys. (1968),Vol.9, 813

- R.Geroch, J.Math.Phys. (1968),Vol.9, 1739

- J.Komorowski, Od liczb zespolonych do tensorów, spinorów, algebr Liego i kwadryk, PWN Warszawa 1978

[111] - M.Atiyah,F.Hirzebruch, Essays on Topology and Related Top. Memoires dedies a G.de Rham, ed.A.Haefliger and R.Narasimhan, Springer Berlin 1970

- C.Chichlinsky, Trans.Am.Math.Soc. (1972),Vol.172, 307

- A.Hattori, Inv.Math. (1978),Vol.48, 7

[112] - P.Forgacs,D.Lüst,G.Zoupanos, in Proc.of the Sec.Hellenic School in High Energy Phys., Corfu 1985

[113] - R.Bott, in Diff.and Combin.Topology, ed.S.S.Cairns, Princeton Univ.Press, princeton 1965, 167

[114] - E.Witten, Nucl.Phys.B (1985),Vol.258, 75

[115] - G.Zoupanos,"Wilson flux breaking and coset space dimensional reduction", CERN-prepr. TH 4830/87 (1987)

[116] - R.Kerner, J.Math.Phys. (1980),Vol.21,No.10, 2553

- R.Kerner, Ann.Inst.H.Poinc. (1981),Vol.34, 437

[117] - G.Chapline,B.Grossmann,"Dimensional reduction example with 3 families of low mass fermions",Lawrence Livermore Nat. Lab. prepr. RU 84/B/86

[118] - CHARMCollaboration, "A precise determination of the electroweak mixing angle from semileptonic neutrino scattering", prepr.CERN EP/86-171 (1986)

notation: ECHAYA - Fizika Elementarnych Chastic i Atomnovo Yadra

ZETF - Zurnal Experimentalnoj i Teoreticheskoj Fiziki

Table 1

No.	G/H	d	K	m_W	m_Z	m_H	$\sin^2 \Theta_W$
1	$\mathbb{C}P^n$	$2n$	$SU(n+2)$	$m \cdot \sqrt{n}$	$m \cdot \sqrt{2(n+1)}$	$m \cdot \sqrt{2(n+1)}$	$\frac{n+2}{2(n+1)}$
2	$\mathbb{C}P^n$	$2n$	$Sp(n+1)$	$m \cdot \sqrt{2n}$	$m \cdot \sqrt{2(n+1)}$	$m \cdot \sqrt{2(n+1)}$	$\frac{1}{n+1}$
3	$\mathbb{C}P^n$	$2n$	F_4 (with $n=3$)	$m \cdot \sqrt{\frac{6}{5}}$	$m \cdot \sqrt{\frac{3}{5}}$	$m \cdot \sqrt{8}$	0.25
4	$\mathbb{C}P^n$	$2n$	G_2 (with $n=1$)	$m \cdot \sqrt{3}$	$2m$	$2m$	0.25
5	$G_{2,n+2}(\mathbb{R})$	$2n$	$SO(n+2)$	$m \cdot \sqrt{n}$	$m \cdot \sqrt{2n}$	$m \cdot \sqrt{2n}$	0.5

with $\mathbb{C}P^n \cong SU(m+1)/(SU(m) \times U(1))$,

$G_{2,m+2}(\mathbb{R}) \cong SO(m+2)/(SO(m) \times SO(2))$.